1903-2003
100 YEARS OF SUSTAINED, POWERED & CONTROLLED FLIGHT

COMPLETELY NEW BOOK

The Good Flight Simmer's Guide Mk.II

Edited by Mike Clark

The Centennial Guide For Computer Aviators

FROM THE PUBLISHERS OF:
WWW.TECPILOT.COM
THE FRIENDLY CLUB FOR FLIGHT SIMMERS

2003

Published by TecPilot Publishing

TecPilot Publishing
23 Pilkington Drive
Clayton-Le-Moors
Accrington, Lancashire,
BB5 5WR, United Kingdom

First Published 2003

© Copyright TecPilot Publishing, 2003
© Copyright Mike Clark, 2003

ISBN 0 954 1967 16

Cover designed by Mike Clark
Printed and bound in the United Kingdom

This publication is copyrighted. No part of this book may be reproduced, stored in a retrieval system, transmitted in any form or by any means, electronic, mechanical, photocopying or otherwise without express written permission of TecPilot Publishing.

Every effort has been made to ensure complete and accurate information concerning the material presented in this book. However, TecPilot Publishing can neither guarantee nor be held legally responsible for any mistakes in printing or inaccurate information contained within this book. The authors always appreciate receiving notice of any errors or misprints.

This book contains trade names and trademarks of several companies. Any mention of these names and trademarks in this book are not intended to either convey endorsement or other associations with this book. They are for reference only.

Microsoft Windows and Microsoft Flight Simulator are either registered trademarks or trademarks of Microsoft Corporation in the United States of America and/or other countries. This product is neither produced nor endorsed Microsoft Corporation or by any other third party.

Introduction

The Good Flight Simmer's Guide Mk.II -2003

It's astonishing to think that it has only been 100 years since the Wright brothers took their first brave steps at Kitty Hawk on the 7th December 1903 and catapulted themselves into the history books by claiming the very first Sustained, Powered and Controlled flight. As you may have seen we commemorate this on this books cover for future generations to see.

100 YEARS OF SUSTAINED, POWERED & CONTROLLED FLIGHT

1903-2003

What is even more astonishing is that over twenty of the centennial years of flight have been spent "flying virtually" - behind personal computers. Yes, over 20% of the total time people have be flying in powered aircraft has also been spent being, "Desktop Pilots" in the home!

Putting this into perspective you could surmise that PC based flight simulation has played an important roll in shaping aviation's modern history. We know, for example, that Microsoft's franchise of Flight Simulator is regularly used as a tool for US forces to train with and that flying schools utilize its navigational features to help students learn the basics of flight. The mind boggles when you think of what will be possible within the next 20 to 30 years.

My point is this; as we mentioned in our first book, flight simulation is MORE than a past-time and people take it very seriously and care a lot about it. It fills a "realistic" niche in peoples' lives that no other PC based software package can offer. It stretches the imagination, like nothing else and it takes users away to places they can only dream of. Flight simulation enthusiasts, rather like the Wright brothers, love a challenge and like to be mentally stimulated. Therefore, in this, our second book, we shall expand on the basics of simulation, learned in our last book and offer more advanced and tailored subject matter without blinding you with science.

We have tried to blend a mix of topics together which cater for everyone's tastes. Obviously we cannot go into enormous detail as we simply haven't got the space to do so. You can always join our club [www.tecpilot.com] where we have unlimited resources at our disposal. One thing we can guarantee though and that is you will make a cup of tea at some point and start asking yourself questions and want to read more. So be prepared.

Lastly, when you have the time and if you are reading this in 2003, please take a moment to stop and think about the hundreds of people who have given their heart, souls and lives to aviation and technology over the past 100 years. The 7 man crew of Columbia for example (which was tragically lost in February of this year) and the Wright brothers who dedicated their lives to giving man perpetual wings a century ago.

Thank you for purchasing this years book. I sincerely hope it helps you accomplish all that you strive for in flight simulation. Next year we shall take you to another level of realism you never thought possible and YES it will be a completely NEW book (like this one) and NOT an update.

Good Reading and Good Flying!

DEDICATION

This book is dedicated to the thousands of TecPilot club members who have supported myself and the team since we started back in July 2001. What you discover in this book is the culmination of months of work. It is based purely on feedback and requests from our much valued online membership. Thanks guys, we couldn't have done it without you!

TecPilot memberships can be obtained by visiting:

HTTP://WWW.TECPILOT.COM

Credits & Acknowledgements

CONTRIBUTORS (In Alphabetical Order)

Andrew Herd
Advanced Hardware
Fluid Frame Rates

Danny Riddell
Build Aircraft, Scenery objects Using GMAX

Keith Davies
Guide to Flying Online

Lesley Clark
Sub Editor, and Moral Support

Mike Clark
All Other Written Material, Artwork, Editor & Publisher

Stacy Barnes-Fullalove
Sub-Editor, Marketing

Steffan Barnes-Fullalove
Sub-Editor, Technical

Stephen Heyworth
Learning to Land
The Big Flight Challenge
Navigating In Winds
Flying An Aircraft To Obtain Specs For Flight Planning
Flaps Explained
Gear Explained
Trimming Explained

IINSPIRATION, INFORMATION, HELP & ADVICE

Adobe (UK) - http://www.adobe.com
AOPA (UK) - http://www.aopa.co.uk
Bob, Paul, Kevin & B&L Distribution Ltd - http://www.bl-distribution.co.uk
Hewlett Packard - http://www.hp.com
Microsoft (USA) - http://www.microsoft.com
Microsoft (UK) - http://www.microsoft.co.uk
Mike Bannister - British Airways - http://ww.britishairways.com
Miguel Blaufuks - The SimFlight Network - http://www.simflight.com
Winfried Diekmann - Aerosoft - http://www.aerosoft.com

Contents

Introduction 5
Credits & Acknowledgements 6

SECTION one

Advanced Hardware 11
  What Kind Of Simmer Am I? 12
  What Kind Of Hardware Do I Need? 13
    Processors - Recommended speeds & types 13
    RAM (Random Access Memory) - Effect on Flight Simulator 14
    Video Cards (Display Boards) - How fast do I go? 14
    Keyboards - Cable V Free 15
    Mice - Cable V Free 15
    USB v Game Ports - What are they and how do they work? 15
    Sound Cards - Wattage, EAX v Dolby - How much is enough? 17
    Speakers - What type do I need and how much should I spend? 17
    Monitors - Seeing Is Believing! 17
    Multiple Monitor Set-Ups - Is seeing more believing more? 18
    Furniture - Desks and Seats. Height makes a difference! 19
  More Advanced Peripheral Devices 19
    Joysticks - A combat gaming device or "all for one"? 19
    Yokes - Or Should That Be 'Yikes'? 20
    Pedals - Does using your feet mean realism? 22
    Throttle Quadrants - I've got to use more parts of my body? 22
    What's Your Perspective? 24
  Time-Out For Reflection! 24
    Radios 25
    AutoPilot Units 25
    Panel Type Add-ons - The Ultimate Experience? 26
  Configuring Hardware Through Menus 27
    Networking 28
    Project Magenta 28
  Conclusion 28

SECTION two

Fluid Frame Rates 31
  Hitting The Silicon 21
  The Processor 32
  Memory 33
  Frame Rates? 33
  Video Cards 34
  Fluidity V Frame Rates 34
  Target Frame Rates 35
  Running Flight Simulator On Marginal Systems 36

SECTION three

Learning to Land 39
  Angle of Attack Versus Speed 40
  What Is a Good Landing? 42
  Speed and Height 42
  Approach to the Basic Landing 43
  The Landing Phase 44
  The Three Point Landing 45
  Crosswind Landings 46
  Reliable Landings 47

Contents

SECTION four

Guide to Flying Online — 49
- The History of Online Flying — 50
- The Zone — 51
- VATSIM & IVAO — 51
- Squawkbox — 52
- Getting Connected — 55
- TACS — 56
- Flight Management System (FMS) — 56
- Virtual Airlines — 57
- Roger Wilco — 57
- Pro-Controller — 57
- Advanced Simulated Radar Client — 58
- Pilot Training — 58
- The Future — 58

SECTION five

Building Aircraft & Scenery Using gMax — 61
- What Is GMAX? — 62
- What Can I Create For Flight Simulator — 62
- How Does It Work — 62
- Is It Hard To Make A Model For Flight Simulator — 62
- Installing GMAX — 63
- The GMAX Interface — 63
- Adding The FS2002 MakeMDL Plug-In For GMAX — 63
- Installing The MakeMDL Plug-In — 63
- What Are Polygons? — 63
- What Are Primitives? — 64
- Adding A Background Reference In GMAX — 64
- Aligning Each Of The Images To Each Other — 64
- Modelling The Fuselage — 65
- Modelling The Wingspan — 66
- Modelling The Engines — 66
- Modelling The Tail Wing — 67
- Naming Your GMAX Objects — 67
- Changing Units In GMAX — 67
- Exporting Your GMAX Model To Flight Simulator — 68
- Creating Your .mdl File In GMAX — 69
- Adding Animation To Your Models — 69
- Re-Using Old Models To Create New Ones — 69
- Texturing Introduction — 70
- Creating A Base Texture In Photoshop (Adobe) — 71
- Creating The Planes Stripe And Airline Name — 71
- Adding Windows, Doors And Metal Panels — 72
- Colouring The Tail Wing — 72
- Colouring The Fuselage — 73
- Painting The Wingspan And Engines — 73
- Applying Textures To A GMAX Model — 73
- Exporting The GMAX File — 74
- Finalising Textures With The Image Tool — 74
- Creating New Texture Types Using The Base Texture — 74
- Creating Basic Scenery In GMAX For Flight Simulator — 76
- Modelling A Simple House Structure — 76
- Modelling A Simple Tower — 76
- Placing New Objects In Flight Simulator — 76

Contents

SECTION six
Navigating In Winds — 79
   Plotting Your Heading — 81
   Flight Plan — 87

SECTION seven
Flying To Obtain Specs For Planning — 89
   Test Flying — 90
   How And What To Test — 90
   Handling Characteristics — 91
   Performance — 91
   Finding V1 — 93
   Landing Performance — 96
   Efficient Performance — 96
   Cruise Performance — 97
   Decent Performance — 98
   Putting It All Together — 98

SECTION eight
Flaps Explained — 101
   A Bit Of Easy Theory — 102
   Design Of Flaps — 103
   When Flaps Are Used — 104
   How To Use Flaps — 105
   The Downside Of Flaps — 105
   Cleaning Up — 106

SECTION nine
Gear Explained — 109
   Tail-Draggers — 110
   Tricycle Undercarriage — 111
   Retractable Undercarriage — 111
   Flying With Retractable Landing Gear — 113

SECTION ten
Trimming Explained — 117
   Trimming Simulators — 118
   How Does Trimming Work — 119
   Why Is Trimming Necessary — 119
   When To Trim — 120
   Why Do Pilots Have Difficulty Trimming? — 120
   Trimming - Final Points — 120

SECTION eleven
The Big Flight Challenge — 123
   Cambridge to Norwich — 124
   Norwich to La Touquet — 125
   La Touquet to Locarno — 127
   The Hardest Flight Begins! — 129
   Big Flight Challenge ICAO Codes — 131

Contents

SECTION twelve

A-Z - Add-On Products — 133
    A Note From Mike Clark (Editor) — 133
    Adventures — 133
    Aircraft — 139
    Scenery — 167
    Utilities — 215

SECTION thirteen

A-Z - Reference — 229
    Web Sites — 230
    Glossary — 236
    Index — 242

SECTION one

Advanced Hardware

Advanced Hardware

What Kind Of Simmer Am I?

Once, long ago and in a universe far, far away, we used Flight Simulator on a Tandy TRS-80. Remember the day we went to the Radio Shack store and stood, holding the box, while we wondered whether to take the plunge? The game cost more than we wanted to pay and demanded a certain amount of imagination - given that you couldn't actually see the plane and the display ran at three frames a second at only 320 x 200 pixels, but we bought it anyway. The fact that there wasn't any panel worth mentioning and that the only way to control the plane was using the keyboard didn't stop Flight Simulator becoming immensely popular; it was such a terrific idea. When Bruce Artwick said, in the 35 page manual accompanying the program, that anyone can beat the three limitations of flying for the price of a microcomputer and the T80-FS1 package,' he can hardly have understood what he was starting.

Nowadays there is such an enormous choice of control systems for Flight Simulator that it is years since we have heard of anyone using the keyboard as their primary control input system. Indeed, Microsoft assume that users will be using a joystick at the very minimum, which has caused a few problems for disabled simmers who have hand-eye coordination difficulties. Where once it was a question of going down to the local computer store and buying the only joystick they had on offer, nowadays someone getting into flight simulation has to decide whether to get an ordinary joystick or a force-feedback one, or perhaps a yoke. Then there is an agonizing decision to be taken about buying a set of pedals, assuming that they will work with the joystick; and over there is a throttle quadrant... The purpose of this chapter is to help users chart a way through the swamp and buy an appropriate control system; and to give more advanced users a picture of the enormous range of hardware add-ons which are available for the hobby.

The first thing you need to do, long before you visit any stores or consult any catalogues, is to ask yourself some questions:

1. **What kind of flight simulation am I into?**
2. **How much time I likely to spend doing it?**
3. **How much am I prepared to pay?**

The reason for asking these three things is that the type of control system you require depends very much upon the answers - the needs of a combat flight simmer being very different to those of a budding 747 captain for example.

Let's go through it. Most likely you already know what kind of simulation you are interested in, if you are reading this book, but the biggest divide is between military sim players and civilian simmers. The categories can be further broken down, as shown in the table on the right.

Military	Civilian
Piston engined fighters	Light General Aviation
Jet Fighters	Corporate Aviation
Helicopters	Airliners
	Jets
	Helicopters

We'll deal with the military/civilian split first, because they have almost diametrically opposing requirements. Military flight simulations are all about speed and manoeuvrability - getting yourself into position to take a shot before fire-walling the throttles and getting out again. Civilian sims, on the other hand, demand balanced turns, precise control of pitch on glide slopes and fine changes in power settings. Clearly, military simmers need control systems which can cope with high rates of roll and rapid alterations in power settings - as well as being robust enough to handle the sort of punishment that is shed out by rough handling in the average on-line dogfight. Combat simmers must have systems which are tough, have large, easily accessible controls, and offer as much independence from the keyboard as possible. By contrast, civilian simmers require more sensitive input systems and need afford to make more use of the keyboard, not only because they have more time to make adjustments to their course and speed, but also because they frequently have to deal with very extended avionics setups.

What if you like to fly Flight Simulator most of the time, but can't resist the odd game of Il-2 Sturmovik? No

Advanced Hardware

problem - if you are prepared to make compromises and select the correct type of joystick, you can use one system to do everything, but as time goes by and your skills develop, you may feel the need to upgrade. By the time you have read to the end of this chapter, you should have a good idea of what path you would like to follow.

Once you have taken some time to work out what kind of flight simulation you enjoy most, the next question to ask is how much time you intend to spend doing it. While the majority of users fly sims very occasionally, the fact that you have bought this book implies that you are thinking of moving up a gear and devoting more time to what many find to be a deeply fascinating hobby. Flight simulation has a habit of taking over large portions of people's lives and while most general aviation (GA) aircraft only have endurance of a few hours, it isn't unknown for simmers who are interested in airliners to spend six or more hours on a single flight - and at the top end of the scale there are people who build facsimiles of airliner cockpits and run networked PC systems driving half a dozen monitors. Many users find the world of flight simulation so fascinating that they move on to reading textbooks and magazines intended for real world aviation in order to broaden their experience and enjoyment of the hobby. Equally, there are many pilots who turn to flight simulation in order to practice various aspects of flying, such as instrument flight conditions, instrument failures, or forced landings.

The amount of interest you have will drive the amount of money you will want to spend on flight simulation, subject to your budget, but as your experience grows it will become clearer how much of your personal budget it is sensible to spend on your hobby. For this reason it is important to make sensible choices when you purchase equipment early on, so that you have the maximum possible opportunity for expansion later. If, for example, you think that you are likely to want to simulate transatlantic 747 flights using 'real' air traffic control and an extended hardware setup, there isn't much point buying a PC that isn't capable of expansion to match that ambition.

What Kind Of Advanced Hardware Do I Need?

Processors
Recommended Speeds & Types

This leads us naturally onto hardware requirements, and because flight simulators are some of the most demanding applications ever written, the first thing to consider is processor speed.

Flight simulators are different to other games because they attempt to simulate an entire environment taking into account physics, aerodynamics, meteorology, radiotelephony, and even human physiology. To give an example, in the apparently simple case of an aircraft taking off from a runway, Microsoft Flight Simulator has to calculate the effects of engine rpm, propeller pitch, fuel mixture setting, flying control incidence, altitude, temperature, wind speed and direction and the loading of the aircraft; the program even simulates ground effect.

It won't surprise you to know that a very complex physics engine is involved, and in general, the coding that lies behind a flight simulator is very different to that of a 'first person shooter' game. Games like Quake only have to make very basic physical calculations, allowing them to devote the majority of their code to optimising display. This explains why a machine which will run Quake III at 80 frames per second (fps) may at times run Flight Simulator 2002 at 15 or 20 fps - because there is a whole different order of magnitude of processing involved. It also explains why flight simulators generally don't demand the fastest possible video cards, because the emphasis is upon simulating the complex nature of real flight, rather than on blazing speed and special effects.

So the answer to 'how fast a processor do I need?' is pretty simple. The bald fact is that you must buy the fastest possible PC you can afford and if a choice has to be made, be prepared to trade video card performance for extra processor cycles. It is always possible to add a faster video card later on, whereas processor upgrades are usually unsatisfactory and rarely cost effective. Although developers publicise their own recommendations, in general terms, few of today's simulators run well on less than a 1 GHz Pentium or the equivalent AMD processor. The editors wouldn't choose to run FS2002 on anything slower than a 1.5 GHz Pentium -

Advanced Hardware

and that is for a basic default setup. If you are considering using complex add-ons with FS2002, such as upgraded international airports, then you should be looking at a 2 GHz Pentium or faster. Readers who are purchasing new machines are in the fortunate position that cut-throat competition among suppliers means that much faster systems are becoming available relatively inexpensive these days.

The rule for upgrading PCs is that speed increases are rarely noticeable unless the new processor is at least one third faster than your existing one and many simmers begin considering an upgrade when PCs with double the clock speed of their existing systems become available. To keep upgrades affordable, the trick is to wait until machines with double your existing PCs specification are no longer the 'leading edge' systems on offer. There is an enormous price premium to be paid for the privilege of owning the fastest possible system and the rate of processor development means that that advantage rarely lasts for very long. So bide your time.

RAM (Random Access Memory)
Effect on Flight Simulator?

The amount of RAM a simulator-optimised PC needs depends to some extent on your operating system and whether you intend to run other applications at the same time. Beyond a certain level, RAM has very little effect on simulator performance and the only advantage is that it provides an overhead for loading additional applications into memory at the same time as the simulator. In practical terms, Windows 9x based systems (including Windows Me) should produce good simulator performance with 256 Mb of RAM fitted and Windows XP based systems are better if 512 Mb is used. After these amounts are satisfied, the law of diminishing returns applies and adding extra RAM generates less and less return in terms of frame rates.

RAM is however, relatively cheap, although as a commodity its price is subject to fluctuations from time to time, and it does not hurt overall system performance to fit 512 Mb as standard, regardless of operating system type. To this end, if you are specifying a system from new and are running close to your budget, it is better to economise on RAM and buy a faster processor, because you can always fit additional memory later on. If you find yourself in this position, an important point to consider is that you don't want to find all the memory slots on your machine filled with RAM when you come to upgrade. Modern PCs use either three or four memory slots, and if possible, it is better to purchase a system with the existing RAM divided between two of those slots than occupying all of them - in the latter case, you will need to completely replace the system RAM in order to upgrade the quantity.

Video Cards (Display Boards)
How Fast Do I Go?

Most flight simmers start with a simple single monitor configuration as supplied by the manufacturer of their PC. In general, such setups are relatively basic because suppliers have little incentive to supply highly specified components, given that most PCs are bought to run office applications. Typically, such an off the shelf setup will include a standard 15" or 17" monitor and basic graphics adapter (frequently built in to the motherboard of the PC) which will display word processor and spreadsheet output without any problems, but is capable of supplying only bare essentials for running games. Manufacturers often reduce costs by supplying rebadged low cost monitors and unsophisticated graphics cards with as little as 8 Mb of RAM, which rules out running many of the special effects that flight simulators are capable of displaying, and may make it impossible to run some add-on packages at all. To take an example, while Microsoft Flight Simulator 2002 will run on a machine with an 8 Mb, DirectX 8.0 compatible 3D video card, doing so rules out using many of the graphics options for which the game is so highly regarded.

So to get the most out of Flight Simulation we recommend a much higher graphics specification. The working minimum can be regarded as a 32 Mb DirectX 8.0 compatible 3D video card. Cards with as much as 128 Mb of RAM are available, but at present few Flight Simulation applications demand more than 32 Mb. Using a 32 Mb card allows the PC to load all the texture files it needs into video memory, with the result that the display is more fluid. Fitting a video card with more than 32 Mb of RAM has little effect on the speed of most simulators at present, although there is little doubt that within a few years, 32 Mb will become the minimum specification, so if you are aiming to 'future-proof' your system, there is an argument for buying a card with as much

Advanced Hardware

memory as possible. Strangely, as far as the majority of flight simulation packages are concerned, video card clock speed is irrelevant, at least as far as the present generation of cards are concerned - there is little point buying the fastest card on offer, because the gain over a mid-range card will be fairly minimal.

This brings us to the knotty question of which video cards are best for flight simulation. From experience, we would recommend the **nVidia GeForce 4** series if you are buying new, as these have been tried and tested with Microsoft Flight Simulator and should run most other sims without any trouble. Feedback from users has produced very few complaints about these cards over the years and they are about as standardised as it is possible to get. However, the hidden strength of nVidia processor based cards is the huge investment the parent company puts into software driver development - new drivers are released on a regular basis on nVidia's website (**www.nVidia.com**). Most simulators will run on GeForce 3 cards without problems. The latest ATI Radeon (**www.ati.com**) graphics cards are also suitable for running flight simulators, but early versions of this series were prone to driver problems.

Looking at specific cards, we would recommend cards based on the **nVidia GeForce 4 MX** for simmers on a budget and the **GeForce 4 Ti 200** series or better for buyers who are looking for the best possible setup.

Keyboards
Cable v Free

A brief word about keyboards and mice is probably sensible at this juncture. If you already have a PC you will be familiar with the rats' nest of cables that a monitor, keyboard, mouse and speakers can leave on a desk and it is worthwhile considering cordless units. The one drawback of cordless input devices is that they are reliant on battery power, but in our experience modern devices are remarkably power efficient and worth serious consideration. Using a cordless keyboard has decided advantages as it frees you from the problem of trying to stretch the cable around the yoke or joystick which inevitably occupies the central position of any flight simulator setup. If you are considering adding extra hardware units, such as radios or a rack mounted autopilot fascia to your system, the cable problem can only get worse.

Mice?
Cable v Free

Much the same goes for mice, although with the added complication that optical mice are widely available and are beginning to replace conventional analogue devices. The advantage of optical mice is that they are extremely precise, which makes them a very good fit with simulators; adjusting altimeter settings or selecting radio frequencies often demands that you should be able to target tiny hot spots. Optical mice have their disadvantages, however, they aren't good if you like to play fast moving games like Quake (if the mouse leaves the desk even for a moment, cursor movement usually stops) and they don't work particularly well on patterned surfaces, though this can be improved by placing a piece of white card under the mouse.

USB vs Game Ports
What are they and how do they work?

The first problem facing any simmer buying additional hardware components such as yokes, pedals or flight panels is how to connect them up to their PC. A few years ago, the answer was simple - almost every device sported a 'game port' connector. **Game Ports** first appeared on sound cards about fifteen years ago and made it possible for a joysticks to become a standard accessory. Nowadays this type of port is becoming less common, but many sound cards still have them and they are occasionally found built in to the motherboard of

Advanced Hardware

PCs. The problem with this ageing standard is that game ports work by translating analogue signals from the joystick into digital ones and the process of metering what the joystick is doing periodically requires the full attention of the CPU, which slows the system right down. Depending on the type of game port your system has, there is a worse snag, which is that there is a limit on how many devices a game port can support. The maximum is four, but doing this usually requires the use of a special adapter to chain the devices together.

Female Game Port connector. This type of connector is commonly found on Sound Cards and Motherboards.

It probably won't surprise you that game port support is decreasing and that modern peripherals are more likely to use a different type of connector, known as **Universal Serial Bus**, or **USB** for short.

USB Female Connector. Found on most "off-the-shelf" computers these days.

The big advantage of **USB** over the game port is that it allows the connection of many more devices but USB is also much faster and more flexible than its predecessor. Theoretically, you can connect as many as 127 different devices to a USB-enabled PC, which should be more than enough for anybody - and the connector can be used to plug in virtually any device you care to imagine, ranging from joysticks to telephones. It is rare for modern PCs to have fewer than a couple of USB ports and the majority have four. In practice, despite the 127 device limit, PCs rarely need to support more than half a dozen USB devices.

You might ask how, given a maximum of four USB ports, it is possible to support over a hundred peripherals, but the answer is that it is possible to add USB 'hubs'. A hub is an external device that plugs into one of the PCs own USB ports - and adds several more. The most common number of ports on a hub is four, but some have six or even eight, and hubs can themselves be chained together to provide dozens of extra ports. One thing to watch out for is that there are two types of hub: powered and un-powered. The reason for this is that the USB standard allows devices to draw whatever power they need from the USB connection itself, which means that un-powered hubs are only suitable if you plan to use a few energy conscious devices.

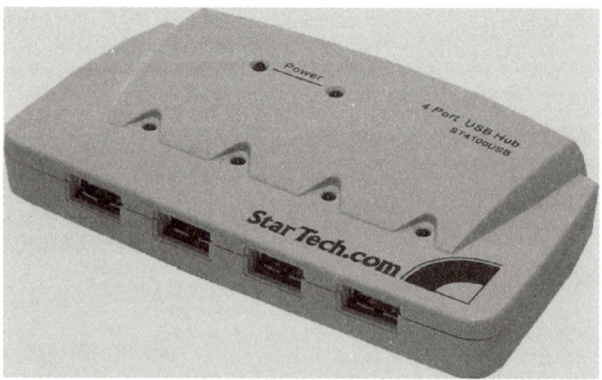

Typical 4-Way USB Hub. Imagine a USB hub like a power adapter or "splitter". Remember that Powered and Un-powered units are available. This example works with onboard USB power or its own adapter. Very handy!

Powered hubs don't cost that much more and run no risk of stressing your PC's power supply, so they are usually the best bet. Perhaps the best thing about USB is that the system uses 'upstream' and 'downstream' connectors that can't be plugged in the wrong way round and they are relatively inexpensive. A new standard called USB 2.0 has just emerged, which is considerably faster than the original USB port and is backwards compatible, but at the time of writing there is no flight simulation hardware which requires USB 2.0 in order to operate.

The reason for going into this level of detail about port types is that hardware conflicts cause a great deal of trouble and flight simulation is the one area of computing where it is still possible to come across peripherals that do NOT conform to modern standards. The reason for this is that it is a relatively small and specialised market, which means that hardware manufacturers can't rely on huge sales volumes to fund development. With the game port fast becoming a legacy device, it is important to have a good understanding of peripheral connection issues BEFORE parting with your money.

Advanced Hardware

Sound Cards
Wattage, EAX v Dolby - How much Is Enough?

Sound is another area where flight simulators differ from other games; the fact is that they don't demand very sophisticated sound systems. Given that speakers are likely to be relatively close to the user, even a 10 watt amplifier can provide plenty of volume and relatively basic sound cards will do; so you don't need to invest in Dolby 5.1 surround sound.

For maximum compatibility, we would recommend a card from the Creative SoundBlaster range. If you want maximum realism, or plan to play other games with more demanding audio requirements, then a card from the Audigy series is ideal. There is a wide range of options, but the cheaper Audigy cards will deliver everything you need. Otherwise, any SoundBlaster compatible card should work as long as it is also DirectSound compatible and there are usually some very attractive deals around. Once again, if you are on a budget, get a cheaper card if it lets you buy a faster processor.

Speakers
What type do I need and how much should I spend?

Although it isn't strictly necessary to spend a great deal of money on a speaker system for flight simulation, better quality boxes do add to the realism. Although a two speaker setup will suffice, it is probably worthwhile investing in a three speaker system with a sub-woofer if you want to enjoy aircraft sound sets to their fullest, particularly if you enjoy flying military sims or helicopters. Another reason for upgrading to a better speaker system is if you plan to install one of the growing number of addon airfields which have their own built in sound sets (for example Lago's award-winning Emma Field) or if you intend to do a lot of flying using a sim which has air traffic control. In each case, staying with a basic two speaker setup will deny you a great deal of enjoyment. Excellent PC speakers are available from Creative, Altec Lansing, Sony and many other suppliers.

The alternative to spending money on speakers is to buy a set of headphones and many simmers choose to do this, particularly if the location of their PC means that they have to limit the amount of stray noise the system produces. A good set of headphones costs relatively little and many PCs have a headphone jack on the front of the CD drive to which a mini jack can be connected; failing this, there is usually an output socket on the back of the sound card. Headphones not only allow you to turn the volume up as much as you like without annoying the family, they are a particularly good way of stopping subwoofer 'boom' which travels effortlessly through walls and is a potent source of complaints from neighbours. There is a vast range of headsets available, ranging from cheap lightweights to hi-fi quality units from Sony and Sennheiser. Cordless headsets are beginning to become available and if your budget allows it, it might be worth considering one, because of the freedom of movement these offer.

> **It is quite possible to enjoy flight simulation using a 15" monitor supplied with a basic home PC system, but sooner or later you will start to notice its shortcomings!**

Monitors
Seeing Is Believing!

It is quite possible to enjoy flight simulation using a 15" monitor supplied with a basic home PC system, but sooner or later you will start to notice its shortcomings. A screen which is perfectly adequate for reading email and composing letters may struggle to display the level of graphical detail needed for flight simulation, which demands large format screens, high contrast and fast refresh rates. Seventeen inches is a sensible minimum if you are considering buying a monitor for flight simulation and nineteen or even twenty-one inch monitors are popular. The reason for using such large displays is because it is essential to display instrument readings legibly and this rules out extended use of 15" and low end 17" cathode ray tube (CRT) monitors. You don't want to be squinting at the screen trying to work out what your airspeed is when you are on 'Short Final' into London Heathrow.

Advanced Hardware

CRT monitors are widely used and have the advantage of being relatively cheap and offering high contrast ratios - which means vivid colours and pure whites and blacks. Monitors which have washed out colour are rarely any good for gaming because details tend to vanish. We mentioned the importance of refresh rates above, the reason being that a screen refresh of less than 75 Hz is tiring after prolonged use. Slow refresh rates can produce noticeable flicker and even unstable images, leading to headaches and other health problems, so it is important to establish whether your video card/monitor combination can produce adequate results. Cheaper monitors often cannot run at high refresh rates and cheaper video cards often cannot deliver them at the kind of resolutions necessary for flight simulation - 1024 x 768 pixels or higher, in full colour.

Recently, liquid crystal display (LCD) or 'flat panel' monitors have become available at reasonable cost, although they are still more expensive than their CRT equivalents. LCDs have the advantage that they require much less space than CRT monitors, have low power consumption, produce little radiation and offer a greater screen area for any given measurement - a 15" flat panel has more visible display than a 15" CRT. On the debit side, flat panels are more expensive and have a tendency to offer less inferior colour and contrast than conventional display units and they are built to display an optimum resolution. Unless you are dealing with the most up-to-date high end LCDs, flat panels also tend to suffer from 'smearing' of moving images, which clearly is undesirable for running games; so if you are thinking of buying a unit, please ensure that you see it in operation before you make a purchase. Another problem with LCDs is that they are prone to pixel 'drop out' and it is not uncommon for units to be delivered - new - with up to half a dozen pixels which will only display a single colour, or white. A unit in this state may well be within the manufacturer's specification for fitness for sale and after purchase it may be impossible to do anything about it. LCDs are at an early stage of development and the rule is caveat emptor - buyer beware.

It is hardly possible to go wrong buying a conventional CRT monitor these days as long as you stick to reputable suppliers such as Sony, NEC, CTX, Ilyama, Philips or Samsung but do check the specification before purchase - there isn't much point buying a monitor that isn't capable of displaying 1280 x 1024 in 32 bit colour at 85 Hz.

> **There isn't much point buying a monitor that isn't capable of displaying 1280 x 1024 in 32 bit colour at 85Hz?**

Buying an LCD, is, as you will have gathered, more complex. This is the first year it has been possible to even make a tentative recommendation that this type of monitor might be used for flight simulation, though once again we would stress that - ideally - you should try before you buy.

The greater visible area an LCD offers means that a 15" unit is just adequate for flight simulation, but we would recommend going to 17" as you will find it more comfortable and as prices drop they are becoming much more affordable. Points to watch are that that the optimum resolution of the panel is at least 1024 x 768 for a 15" and 1280 x 1024 for a 17"; that the brightness is at least 250 cd/m2 and preferably 300 cd/m2 and that the contrast ratio is at least 300:1 and preferably higher. One more thing to check is the type of connector on the LCD - the better ones offer digital as well as conventional analogue inputs - which means that if your graphics card has digital output, you will be able to get the best possible image on your LCD.

Multiple Monitor Set-Ups
Is Seeing More Believing More?

Some flight simulation packages (such as Microsoft FS2004) allow the use of multiple monitors to display screen output. We will tackle this area in greater detail later in this chapter but if your video card is capable of driving two monitors, it is possible to use the second display to show additional views, although this is generally at the expense of overall speed. So if you are thinking of buying an LCD panel, don't dispose of your old CRT monitor, because it may be possible to use it to extend your flight simulation experience. Similarly, if you are buying a new video card, it is worth checking to see if your choice has multiple monitor support available. More modern operating systems such as Microsoft Windows XP have dual monitor support built in and this is remarkably easy to configure and use.

Advanced Hardware

Furniture
Desks and Seats. Height makes a difference!

We make no apologies for including a section on furniture here, because lack of attention in this area can spoil enjoyment of flight simulation and is also a source of troublesome musculoskeletal problems. It is absolutely essential to buy the best quality desk and chair that you can afford if you want to get the most out of your sim. A great deal of this is basic common sense but we hear of many simmers spending hours, perched on inadequate chairs, squinting at badly sited monitors, who then wonder why they suffer from chronic shoulder and neck ache. Repetitive strain injury springs to mind, doesn't it?

When you select a chair, make sure that it offers good back support; ideally it should be adjustable for height as well. Swivel chairs can be used with flight simulators, but they aren't ideal because they tend to move at the wrong moment - there is nothing worse than watching your aiming point creeping towards your enemy's empennage only to have your chair skate away at the critical moment! If you do use a swivel chair, placing it on a thick piece of carpet can reduce unwanted mobility.

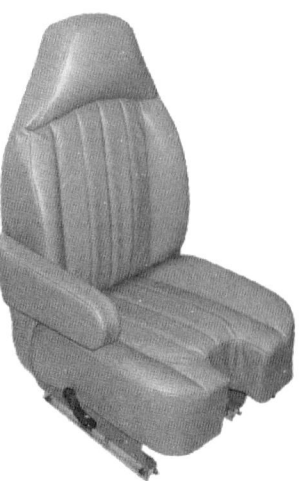

PFC's Captain's Seat

If you want to go all the way, PFC offer a 'Captain's' seat in leather or vinyl, which is a unit actually designed for a commercial aircraft. This has an adjustable base and a reclining back and can be purchased with optional seat tracks, so it can be fixed to the office floor.

Desks introduce a different problem, in that a desk which is the correct height for typing and monitor positioning is usually too high for comfortable mounting of a yoke or joystick. This leads to an unnaturally high position for the control unit but unless you cut down the legs of an old desk or have one that is adjustable for height, this is a difficult problem to solve. It is essential, however, to ensure that the desk is sited so that the monitor faces you and is shaded from direct sunlight or reflections from nearby windows.

More Advanced Peripheral Devices

Joysticks
A combat gaming device or "all for one"?

The joystick is such a ubiquitous piece of hardware that the vast majority of gamers probably already own one, but flight simulation poses its own special requirements. That being said, it is quite possible to enjoy virtual flight with a basic joystick because the majority of simulators allow a considerable degree of keyboard control. However, as your experience grows, you will discover that using a keyboard to control aircraft throttles and flaps isn't particularly intuitive and most simmers graduate to more advanced joysticks fairly quickly. Good quality basic joysticks are available from Microsoft, Logitech and CH Products, but virtually any gaming store will offer a wide selection.

CH Products' Fighter Stick (USB)

It isn't necessary to spend a fortune on a joystick as long as you have a firm idea of what you want it to do. At the most basic level, the reason you need a joystick at all is to allow you to control pitch and roll in other words, to climb or descend, and to bank into turns. Given that most simulators have an option for 'auto-rudder' which produces coordinated turns without the need for pedals, it is possible to do without the ability to control yaw (side to side movement of the nose), but fortunately many better quality joysticks allow you to yaw the plane by twisting the joystick handle from side to side.

The Good Flight Simmer's Guide Mk.II - 2003   19

Advanced Hardware

Other features required of a medium level joystick are a hat control so that you can swivel the pilot's point of view (POV), a throttle control so that you can fine tune power settings, and three or four extra buttons to which you can allocate functions like gear and flap retraction - and a trigger if you are a military simmer. Cordless joysticks are available and make a great deal sense if you are fighting with lots of other peripherals on the same desk.

The most sophisticated joysticks offer a bewildering range of functions, 'force feedback' being a good example. The problem with basic joysticks is that they have little built in 'feel' beyond that provided by spring systems. Not only does lack of feel make it quite hard to fly a virtual plane, it can even be tricky working out when it has landed! Force feedback introduces a new level of realism in simulators which support it, by reproducing control forces, vibration and shocks. So, for example, a force feedback joystick might give a realistic impression of the way controls stiffen at higher speeds, of rumble as the gear touches the ground, the 'bump' as wheels retract, and the juddering of cannons firing. At the end of the day the control forces on a real plane are like, but it does give a much better impression than a passive joystick can. Private Pilots reading this chapter might care to take note that the big difference between real aircraft joysticks and gaming ones is the 'throw'; gaming joysticks have a lever many times shorter than the sticks in light aircraft and it takes time to get used to it. Force feedback joysticks are available from Microsoft, Logitech, Thrustmaster and several other suppliers and they provide reasonable simulation of the controls found in modern military aircraft and even civilian jets such as the Airbus.

The now legendary SideWinder Force-Feedback Joystick from Microsoft.

Force feedback is not the last word in joystick design and some high end developers have avoided it, relying instead on using high quality components with generous internal springing to supply appropriate control forces. Thrustmaster's HOTAS Cougar, is an excellent example - this is a replica of an F-16 throttle and stick and it is built out of metal like the original. The Cougar has eighteen pound springs, adjustable resistances, 28 buttons, multiple hat switches and flash upgradeable memory to allow for compatibility with future games and operating systems. While the Cougar is designed for use with military simulators - the USAF uses it for pilot training - it has a place with civilian sims and is widely regarded as one of the best joysticks on the market.

Thrustmaster's HOTAS Cougar. A replica of a real F-16's Joystick.

Yokes
Or Should That Be 'Yikes'?

Although joysticks are versatile and are the best flight control option for fans of military sims, they don't provide the whole answer if your main interest is in general aviation or commercial flight simulation, purely and simply because the majority of civilian aircraft use control yokes. There are many exceptions to this rule and many people own a share aircraft which utilise sticks, but nonetheless most modern civilian aircraft are flown using a yoke and many Private Pilots have never used a stick. Yokes combine the functionality of a car steering wheel to control roll with an additional fore and aft axis that controls pitch; and while most people find them instinctive to use, they have the disadvantage of precluding rapid control deflections, which is why aerobatic and military aircraft don't use them. The advantage of a control yoke in a light plane is that it clears the floor and makes cockpit access much easier - this isn't an issue in aerobatic or military aircraft.

Advanced Hardware

Yokes have few uses in the gaming world outside flight simulation, so they are only available from a few specialist manufacturers. The most popular yokes are available from CH Products, who make both GamePort and USB compatible devices. Most pilots have used their Flightsim Yoke for years and find it perfectly adequate, even though it looks fragile and is built almost entirely from plastic. There are two versions: one with three control levers on top of the casing and another with only one lever, the latter being known as the Flightsim Yoke LE. While the LE version is cheaper, it makes mixture and propeller pitch control more difficult as those functions can be allocated to the additional levers on the three lever version. In addition to the yoke and levers, both products have hat switches, a trim wheel, two rocker switches, four push buttons and two flip switches which can be used to operate the flaps and undercarriage. Unless you have particularly high standards, the 3 lever Flightsim Yoke is perfectly adequate as a control system and will provide years of service with little signs of wear. The disadvantages of this yoke are that it uses relatively weak (and noisy) internal springs for feedback and the trim wheel range is soon exhausted, given that it only turns through approximately forty five degrees - unlike the trims on real planes, which have much wider excursion.

CH Products' Flight Sim Yoke (USB)

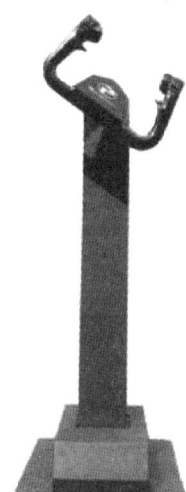

AFCS III USB High Quality Yoke

There is a variety of metal yokes available, although these are much more expensive than the CH Products devices. Vmax market the **AFCS III USB Advanced Flight Control System**, which is based on a SAAB 340 turboprop yoke moulding. This is a very high quality product, based on a heavy steel base built of steel and aluminium and it comprises of a yoke and throttle. In addition to the throttle, the unit has eight switches. Its chief weakness being the lack of a hat control, but few (if any) metal yokes have one. The AFCS III conveys the feel of a real aircraft yoke better than a plastic device, even though it does not feature force feedback.

The Cirrus series of metal yokes are available from Precision Flight Controls (**www.flypfc.com**), a major player in the flight simulation hardware field. There are three styles: the Jet yoke; the Mooney yoke; and the Beech yoke and they differ mainly in shape and number of switches. All are available with USB plugs and the Beech yoke even has an option for fitting a digital clock/timer within a central bezel. Once again, none of the yokes have hat controls, though they all have push-pull throttles.

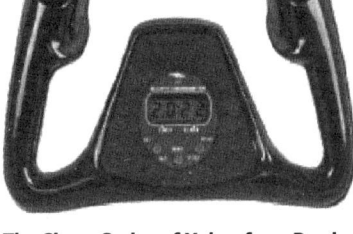

The Cirrus Series of Yokes from Precision Flight Controls. Made from Solid Aluminium.

Precision Flight Controls retail a **'Jetliner Yoke'** which is a full size floor mounted stand alone unit that does away with the need to clamp the yoke onto a desk but this is only for the truly dedicated simmer, since it is a large unit that virtually demands a permanent installation and commands a high price - but there isn't anything else quite like it.

Precision Flight Controls' "Jetliner Yoke"

The Good Flight Simmer's Guide Mk.II - 2003   21

Advanced Hardware

CH Products' Flight Sim Pedals (USB)

Pedals
Does using your feet mean realism?

The biggest step many flight simulation enthusiasts take towards improving the realism of their control setup is purchasing a set of pedals. Although most simulators feature 'automatic rudder' systems which allow balanced turns to be made using a stick alone, without pedals it isn't possible to intentionally skid a plane, as a fighter pilot might do in order to throw an enemy off his aim, or a civilian pilot might when using the wing-down method to make a cross-wing approach. In broad terms, simmers who are mainly interested in civilian aircraft have the least need for a set of pedals, while it is hard to imagine flying a realistic combat without them but every user stands to gain from using pedals. The twist grips found on some joysticks do give a certain amount of rudder control but it is far less instinctive and definitely less realistic than controlling the yaw with your feet.

The best example of a moment when rudder control is essential is during the last few moments after the flare, as an aircraft settles onto the runway in a strong crosswind - using an auto-rudder feature, the only way to control the drift is to bank the plane, which is impractical, unrealistic and hardly an option if you are flying a 747.

Once again, the beginner's standard for flight simulation pedals is made by CH Products. Built from plastic, they are a remarkably sophisticated three axis product and given the correct software setup it is possible to use differential toe braking as well as controlling the rudder axis. Both game port and USB versions are available.

Moving up in quality terms, Precision Flight Controls offer a wide range of metal rudder systems, albeit at a substantially higher price than the CH Product set. Single and dual pedal systems are available in designs which suit both GA and airliner scenarios and all the PFC pedal sets feature toe braking. At the time of writing USB versions were not available but they are said to be under development. The advantage of the PFC pedals is that they are much heavier and harder wearing than the CH Product design.

Cirrus Pedals from Precision Flight Controls. Later to include Proportional Breaking.

Throttle Quadrants
I've got to use more parts of my body?

If you opt to buy a metal yoke and pedals or want greater operational realism for multi-engine aircraft, the next stage is to buy a throttle quadrant. With the exception of the ThrustMaster Hotas Cougar, the throttle systems supplied as standard with joysticks and yokes are rarely realistic and offer minimal features.

At the cheaper end of the market there is a variety of throttles available, although these tend to be aimed at the military simulation market. At present the best example is the CH Products Pro Throttle USB, which in addition to offering a jet fighter style throttle slide, has four hat switches, a three way mode switch, three on/off

Advanced Hardware

buttons and a mini joystick control. Thrustmaster offer a rather more robust metal unit as part of the Hotas Cougar package.

CH Products' Pro Throttle

PFC's Throttle Quadrant Console

Until recently the choice of GA throttle units was severely limited (there weren't any) but Precision Flight Controls have filled the breach with their excellent Throttle Quadrant Console. This is a sophisticated nine quadrant system which is compatible with a wide range of flight simulation packages ranging from FS2002 to Elite and it is based on an extremely clever idea. The base unit features landing gear and flap switches and a rudder trim and a central slot into which different throttle quadrants can be plugged. The choice of quadrants extends from a simple two lever single engine unit all the way up to a four engine jet control set. There are several different twin engine sets ranging from a Baron-type unit to a six-lever turboprop and there is a 737 style three engine jet fitting. Throttle quadrants are purchased separately to the base unit and changing units is a simple matter of undoing two thumb screws and swapping them around. The entire system is manufactured from sheet metal and bar the fact that it still relies on a serial port connection there is no doubt that it will stand the test of time.

The Robust Hotas Cougar Throttle by Thrustmaster

One of the most interesting things about the range of stand-alone throttles on offer is that there isn't one (that we are aware of) which simulates the Cessna-style push-me pull-you type of carburettor heat/throttle/mixture controls that are fitted on approximately 50% of real aircraft and which are familiar to generations of flight simmers thanks to Microsoft's choice of panels. Instead, suppliers have chosen quadrant type units - why? It is hard to guess.

At the top end of the market, SimWare (**www.simw.com**) are offering an impressive, if seriously expensive, two engine jet throttle control unit which connects to a PC via the printer port and duplicates many of the functions of a 737 engine console. This is a sizable unit measuring 30 x 30 x 65 cm and it has two thrust, starter, flap and spoiler levers, a trim wheel, parking brake and warning lights, two TOGA pushbuttons and two auto throttle disconnect buttons - it is hard to think of a more sophisticated unit, but given that you could take many flying lessons for the price of the unit, it remains to be seen exactly how many sales it will make.

Seriously expensive (but wonderful) two engined jet throttle control unit available through SimWare. How professional do you want to be? Well, with this you'll be safe in the knowledge that you're flying your 737 with the right piece of kit!

The Good Flight Simmer's Guide Mk.II - 2003   23

Advanced Hardware

What's Your Perspective?

"Point Of View" (POV) control is a big issue for some flight simmers, given that hardware hat controls do a poor job of duplicating the freedom of gaze open to a real pilot. Various solutions have been proposed to solve this problem, but one of the most ingenious is a device by eDimensional (**www.edimentional.com**) called TrackIR.

TrackIR controls view scrolling by monitoring a dot placed on the user's forehead or glasses. It works with other games besides Flight Simulators and is remarkably simply to set up. It consists of a Cyclops's eye that perches on top of the computer monitor and connects to the PC via a USB cable. The 'eye' monitors the user's head movements and can be used to pan around a virtual cockpit, select external views, or even to change weapons (in combat sims).

TrackIR, by eDimentional. A new "Look" at viewing virtual aviation.

eDimensional also market a set of Wireless **E-D** glasses (**ED** - as in eDimentional) which can be used to provide a 3D appearance with various games. It should be pointed out however that these do not work with LCD monitors at present.

Finally, various touch pads are available, such as Aerosoft's popular Flightboard 2000 USB (Updated since our last book), which allows programmable selection of up to 128 different functions via a touchpad. This kind of device is invaluable for coping with complex cockpit setups.

eDimentional wireless glasses. An alternative to those weird looking forehead Dots! What would you prefer to try captain? Shades or dots?

Aerosoft's Flightboard 2000 USB

Time-Out For Reflection!

At one time, if you had bought a yoke, pedals and throttle unit, you would have reached the end of the road as far as simulation hardware was concerned but nowadays the sky IS the limit. Flight simulation has become so popular that extended hardware setups are becoming available, some of which have price tags which make one wonder if it might not be more sensible to buy a real aeroplane. We make this point advisedly; in the UK, at today's prices, it should be possible to gain a National Private Pilot's License for about £4000 and a JAR PPL for about £5000. Shares in aircraft are available for anything from £3000 upwards and it is possible to buy a Cessna 152 for £15000, so there is an upper limit to how much it is sensible to spend on simulating flight - when the alternative is actually doing it. Of course there are many reasons why people use simulators rather than going for a real flight and the one thing which weighs heavily in favour of simulators is that once you have built one, your expenses are minimal and bad weather can never ground you.

Though comparatively few simmers are likely to spend large sums of money buying sophisticated items of hardware, the following sections introduce even more advanced goodies: Radios, Autopilots and Panel-Type add-ons. These are popular purchases, believe it or not, and ones which certainly aren't for so called "gamers".

If you spend many hours a week using a flight simulator, this kind of hardware definitely adds to the enjoyment and specialist hardware suppliers have developed stylish systems which are popular with flying schools. In some jurisdictions (not many), simulator time can even be accredited towards licenses. Simulators are ideal for practicing flight in IFR conditions safely and several have been developed precisely for this purpose.

Advanced Hardware

Radios

Itra's ActivePanel Radio

A number of companies make radio panels suitable for use with flight simulators. Itra's (**www.sim.itra.de**) ActivePanel Radio is a USB device which is compatible with later versions of Microsoft Flight Simulator. It interfaces with the sim using a gauge which is added to software panels via an installation routine and once installed, allows remote setting of both Com and Nav radios, as well as the ADF. It is also possible to tune VOR radials using the panel.

Precision Flight Controls market an even more sophisticated unit which duplicates one of the old Bendix King "K" type stacks which are still found in many aircraft. Their avionics stack has everything from pairs of flip-flop Com and Nav radios, through an ADF, transponder and autopilot and it is hard to imagine anything that has been left out - in fact I have seen very few real aircraft with such a sophisticated avionics setup. Elite offer similar units as part of their excellent AP-2000 and AP-3000 Avionics panels and yet another is available as part of the Real Cockpit company's RS 372 Bendix King radio stack.

Aerosoft market the popular ACP Compact (as mentioned in our last book), an extremely neat and remarkably cost effective little box which incorporates dual Nav and Com radios, an ADF, autopilot and various other items including a gear control. Ideal for simmers who are short on space, this USB unit can be clamped onto the side of a computer monitor, leaving the desk uncluttered but some users find the controls a little small and too fiddly for comfort.

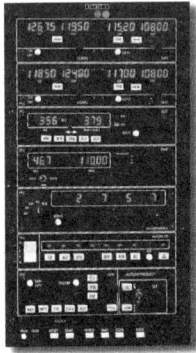

PFC's Avionics Stack

If you want MORE choice, you are on a budget, or fancy a modular approach, GoFlight (**www.goflightinc.com**) may have the answer in their GF-AC cockpit control system, which allows the user to build up an avionics rack at his or her own pace. The base of the system is a mounting unit into which a wide variety of USB devices can be connected and various radio panels are available. GoFlight: "With the GF-AC cockpit control system you have over 400 cockpit commands at your disposal. Flip real switches to start engines. Push real buttons to engage and disengage autopilot. Turn real knobs to tune your radios, and see the frequencies on real LED displays."

GoFlight's AC Cockpit system

AutoPilot Units

There is less choice of autopilot units but nonetheless, the quality of those that are on offer is excellent. Once again, Itra sell a commercial jet mode control panel styled ActivePanel Autopilot unit which connects via USB and works with recent versions of Microsoft Flight Simulator. Like their radio panel, this is a well engineered piece of kit that is designed to do long service. GoFlight manufacture the GF-MCP, an autopilot module that is, once again, optimised for simmers who enjoy flying big jets. As noted above, PFC market a Bendix King autopilot as part of their avionics stack, this being one of three GA style hardware autopilot that we are aware of - the others being sold as part of the Elite and Real Cockpit Ranges of Panels.

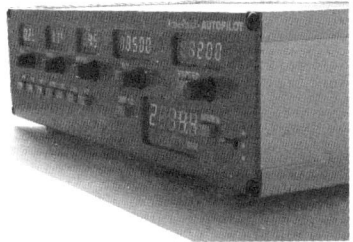

Itra's ActivePanel Autopilot

For the serious hobbyist, PFC (**www.flypfc.com**) also market a wide range of switches, display units and other multipurpose flight controls, so if you are considering building a home simulator, their website is well worth a visit.

The Good Flight Simmer's Guide Mk.II - 2003   25

Advanced Hardware

Panel Type Add-ons - The Ultimate Experience?

If hardware radios and autopilots aren't enough, then the sky is the limit. Precision Flight Controls market a Jet Cockpit Glass Trainer which is effectively a complete aircraft nose unit in a box - a BIG box. This unit (right) is large enough to sit two people inside and comes complete with everything from an overhead switch panel to dual flight controls and glass instruments. For the less ambitious, the same company markets a remarkably wide range of hardware consoles featuring yokes, throttles, gear levers, gauges and switch sets. These range from single yoke consoles, through a dual yoke setup, to a jet airliner pedestal which can be configured with two, three or four engined throttle quadrants.

PFC's Jet Cockpit Glass Trainer

Elite's single yoke flight control panels

Elite (**www.flyelite.ch**) manufacture a pair of single yoke flight control panels, but the prize for the MOST sophisticated GA flight panels must go to The Real Cockpit company (**www.therealcockpit.com**), who produce the most fantastic set of Cessna style light aircraft hardware. (See below)

The TRC372 and TRC572 units are based on the 172 cockpit layout and a remarkable likeness to the real thing. Both units connect to a PC via a single USB connection, the difference being that the 372 is desk mounted and requires a yoke and pedals, while the 572 is a free standing full panel including dual yokes, pedals and a sophisticated avionics stack. The Real Cockpit company plan to release an even more sophisticated unit featuring a full motion platform at a later date.

The Real Cockpit's Panels. As they say in their own marketing material, "All instruments and gauges are accurate reproductions of the originals, without any concessions! "

Advanced Hardware

Configuring Hardware Through Menus

One of the problems with using dedicated flight simulation hardware is finding a way through the maze of different software control settings in order to produce realistic 'feel' and control outputs. The trouble is that three layers are involved in the feedback loop. The first being the assumptions made by the designer of the flight simulation program. The second being the way the flight dynamics of the plane have been loaded by the user. The third being the setup of whichever control device the user happens to favour. Clearly, given half a dozen simulators, several hundred different control devices and tens of thousands of flight simulator add-on planes, it isn't possible to give more than general advice about how to set a home simulator up to give maximum possible realism. However, to give an idea of what can be done, we will briefly look at Microsoft Flight Simulator 2002. Similar options are available in FS2004, however at the time of writing this FS2004 is still in the final testing stages. We'll cover specific software performance issues in the next chapter entitled "Fluid Frame Rates".

Given that there isn't much an end-user can do to alter the flight dynamics of add-on planes, there are limitations on what can be done with a control setup, because flight modellers use such a wide variety of control systems. To give an example, a developer who designs a simulation using a stiff yoke that requires considerable control force to generate roll is likely to produce a sim which feels extremely lively to someone who flies it using a standard joystick. The problem can be ameliorated by visiting the **options/controls/sensitivities** page from the FS2002 top menu and pulling the relevant sliders towards the left, so that it takes larger control inputs to produce a given degree of roll - but while this will fix the trouble with the plane in question, it may well make others completely un-flyable. While on the subject of the sensitivity settings, it is worthwhile setting 'null zones' to about the levels shown, as this improves the rather squirrelly handling that is such a marked feature of FS2002 planes.

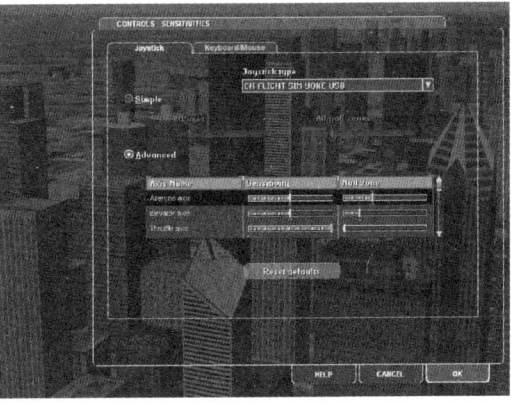

Sensitivities Options In Flight Simulator

The other way of improving handling is to use a different control device. This is a matter of trial and error - for example, it isn't that easy to fly a jet fighter using a yoke, because the device isn't suited to the plane. Fortunately, it is reasonably easy to change control systems using the **options/controls/assignments** menu and Windows control panel joystick set-up applet and it is even possible to have a joystick and a yoke installed simultaneously, although this can lead to weird events when the operating system 'forgets' which device it is supposed to be talking to. The usual symptom being loss of control surface command with the 'active' input device.

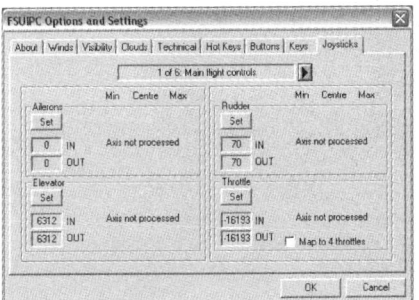

Pete Dowson's FSUIPC utility

Beginners will probably find it easier to fly Flight Simulator with the P-factor slider pulled all the way to the left, because this makes it far easier to control piston engined aircraft in level flight and if you find it difficult to sideslip a plane, check that you didn't engage the yaw-dampers on the last jet you flew, because this parameter has a habit of remaining engaged. Finally, if you are having trouble controlling a plane and think your hardware might be at fault, download a copy of Pete Dowson's award-winning FSUIPC utility (**www.schiratti.com/dowson.html**) which has an option for mapping errant joystick axes.

The Good Flight Simmer's Guide Mk.II - 2003

Advanced Hardware

Networking

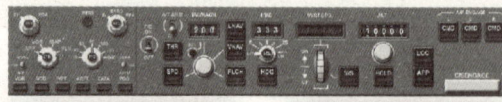

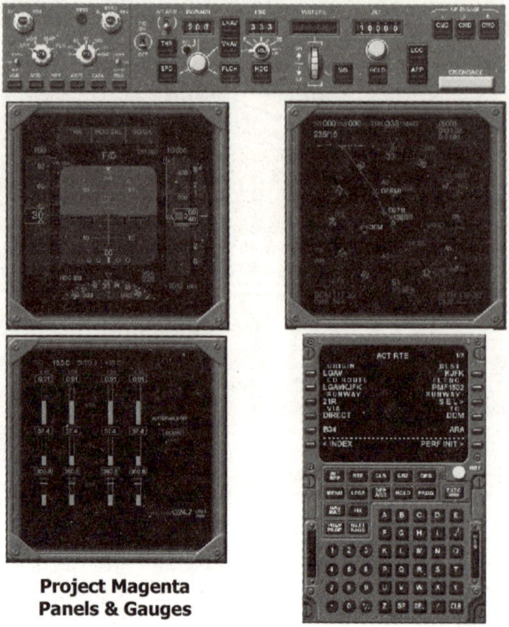

By now you will have gathered that as far as hardware is concerned, the sky is the limit for flight simulation. Finance permitting, there is no reason why you should limit yourself to a single PC/monitor setup and if you download a copy of Luciano Napolitano's WideView (**www.wideview.it**) for Microsoft Flight Simulator, you can create a multi-monitor virtual cockpit using several PCs, each of which shows a different view from your aircraft's cockpit. This very clever piece of freeware solves the Flight Simulator multi-monitor problem - which is that running extra views generated by the same PC can slow the system down to around five frames per second. A WideView based system can be used to run as many different views as you like, providing surround vision, if desired.

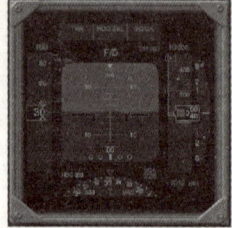

Project Magenta

Project Magenta Panels & Gauges

Perhaps the ultimate simulation challenge is to build a hardware setup to run Project Magenta (**www.projectmagenta.com**). Project Magenta offers a variety of software products which allow the end user to build a jet airliner cockpit based around networked PCs each of which is dedicated to running a single instrument. In this way, a 'glass cockpit' can be built up with life sized gauges running in real time within a realistic panel - but the snag is that you have to build and pay for the rest of the simulator. The challenges involved haven't deterred ambitious simmers from building what are some of the most impressive home flight simulation set-ups on the planet. Visit the web page to get an idea of what is on offer.

Conclusion

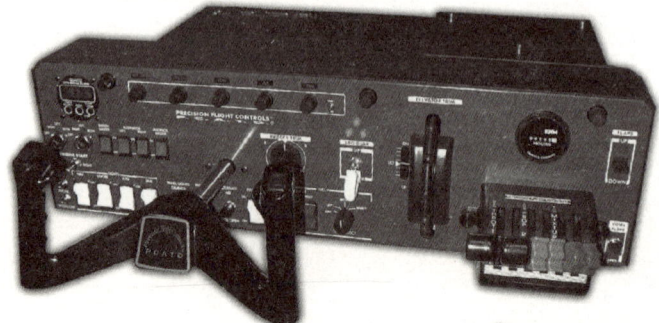

Precision Flight Controls Single Yoke Console

We hope this chapter has given you a comprehensive overview of the current state of hardware in relation to flight simulation - and helped you expand on the knowledge you gathered in our last book (The Good Flight Simmer's Guide 2002). If your previous impressions have been based on our last book OR a quick trip around your local hardware store it really is time to think again - there is a wide range of specialist hardware suppliers out there who are prepared to sell you anything from a simple 3-axis, non feedback joystick to a fully fledged jet cockpit with all the bells and whistles. The only limit is your imagination... and your bank balance of course!

by Andrew Herd
For TecPilot

COMPUTER Pilot
...the Magazine for Desktop Pilots and Flight Simulator enthusiasts

It's The Flight Simulator Pilot's Magazine!

Start receiving the best add-on available to enhance your passion for flying at your desktop...

Computer Pilot Magazine.

Computer Pilot comes to you <u>monthly</u> in spectacular full color throughout and contains nothing but flight simulation... page.... after page.... after page.

It's read by tens of thousands of flight simulator enthusiasts worldwide as their <u>definitive</u> flight simulator reference publication.

Computer Pilot will enhance your enjoyment of flying at your desktop by making you a better, and more knowledge flight simulator pilot.

But why take our word for it... here's what our readers have to say:

"I recently received a trial issue to **Computer Pilot** and when I opened it I was enthused. I have never seen so much information about every flight sim on the market. I am just amazed with how much you can pack into this magazine."

"Of all the flight sim mags I have read, this has got to be the daddy of them all!"

For more information or to subscribe right now go to:

www.computerpilot.com

SECTION two

Fluid **Frame Rates**

Fluid Frame Rates

When Flight Simulator 2002 was released, the first thing most people noticed was how smooth it was compared to FS2000. The juddering and sudden slow downs that had cursed the earlier version of the sim were gone. it was fluid. No longer were turns interrupted due to scenery reloads. No longer were there any unexplained pauses. No longer did everything slow to a crawl on the approach to a complex airport. Online forums were full of messages congratulating Microsoft on achieving what was popularly thought to be impossible; a version of that ran well on affordable PCs.

Even before the first flush of enthusiasm for the product wore off, there were a few dissenting voices. People complained of jerkiness in spot plane view, stuttering in mountainous areas, blurry textures and dramatic frame rate drops. Clearly there were still some problems but the question was; how could it be that two users, sometimes with apparently identical PCs, could have such different experiences? This article explains why.

Hitting the Silicon

The reason Flight Simulator is so processor dependent is largely due to the sophisticated physics engine which lies at the heart of the sim. This calculates everything from the effect of engine rpm and the bank angle on the flight path, through the effects of gusts and turbulence, to such minutiae as spool up times on jet engines. Add in several AI aircraft and a windy day and the calculations become an order of magnitude more complex. This isn't all the sim has to do, it has to calculate the position of mesh and airport scenery, load textures, AutoGen and display the visual model of the aircraft and that is before applying any effects. It isn't surprising that less powerful machines struggle.

The Processor

The best place to start is with Microsoft's minimum spec for Flight Simulator, which is:

- **A multimedia PC with Pentium II 300Mhz equivalent or higher processor.**
- **8MB 3D video card supporting DirectX 8.0a - or better.**
- **64 MB RAM.**
- **650 MB of hard disk space, with an additional 100MB for the swap file.**
- **Quad Speed or better CD-ROM drive.**
- **Super VGA monitor supporting 800x600 in 16 bit colour or better.**
- **DirectX 8.0a API or later installed.**
- **DirectX8.0a API compatible sound card with speakers or headphones.**
- **Microsoft Windows 98, Millennium Edition, 2000, or XP operating system.**
- **Flight Simulator is not supported under Windows NT.**
- **Joystick or flight yoke.**
- **Microsoft Mouse or compatible pointing device.**

Even though this is a minimum spec, some of it is negotiable. For example, Flight Simulator will run on a slower processor but it struggles. Using a low end machine will mean pulling the display settings sliders back to their stops, which defeats the object of the exercise of buying Flight Simulator in the first place. The bottom line is that if your PC is slower than 300 MHz and you can't afford to upgrade, you would be better off staying with older, now outdated versions like FS2000 or even FS98.

So the first point to make about Flight Simulator and about FS in general is that it is "processor bound". The faster the processor, the better Flight Simulator will run. As a rule of thumb, about 60% of 's performance is dependent on processor speed - if your PC hasn't got the horsepower, upgrading the RAM and the video card will make little difference.

Fluid Frame Rates

Following from this, the inevitable question is, how fast is fast enough? Well, experience tells us that the system spec for getting an acceptable performance from Flight Simulator is a rather better machine than Microsoft suggests, something like a 600 MHz Pentium with 128 Mb of RAM and a 16 Mb video card. Even this spec won't run the sim at its best, although some people report being quite happy with it on slower machines. In fact Flight Simulator doesn't begin to come into its own until it is running on a 800-1000 MHz machine and it isn't until it is installed on 1.4 GHz machine or better that some or all of the display sliders can be set to their limits. This is a power hungry programme.

Memory

RAM is less important than processor speed, accounting for perhaps 10% of the performance equation. The floor is very definitely 64 Mb - any less than that and Flight Simulator won't run. Increasing RAM helps, to a point; you will see small but noticeable improvements in fluidity and texture handling up to 512 Mb of RAM. Above that level adding more doesn't seem to make any difference and the performance increase between 256 Mb and 512 Mb of RAM is relatively small.

Frame Rates?

The frame rate is the number of times per second Flight Simulator repaints the image you see on the screen. Frame rates are normally given as frames per second, abbreviated to fps. You can display the frame rate by hitting [SHIFT] and [Z] twice, which will pop up a figure at top right of the display while Flight Simulator is running. Try flying around with the frame rate displayed and see how it changes as you fly from blue sky into cloud, or approach 'Meigs' from over the sea. Note how the frame rate rises with altitude and that it varies with the different views in Flight Simulator. If you want to get rid of the frame rate display, hit [SHIFT] and [Z] again until it clears.

Fluid Frame Rates

Video Cards

For all the furore about video cards, they make less difference to Flight Simulator than you think. As long as your card has adequate memory and has 3D capabilities, its effect on overall fluidity is small, perhaps as little as 5%. This sounds surprising but the display requirements of Flight Simulator are modest compared to games like Quake and for that reason, if you are buying a video card purely to run , there isn't much point getting anything better than a GeForce 3. There are a few snags in this area, though. nVidia card owners have very few problems with Flight Simulator, which is a reflection of the quality of the drivers the manufacturer supplies. ATI Rage owners, on the other hand, reported many problems early after Flight Simulator was released, though these may well have been eased by the latest drivers. The sim runs badly on Voodoo cards, with problems ranging from a complete failure to load aircraft textures to complete lock ups. Given that the Voodoo range is not going to be developed any further, if you have one of these cards it might be worth considering a replacement.

Fluidity V Frame Rates

Having dealt with the hardware, let's consider how the flight simulator software setup can affect fluidity. The first and most important thing to realise is that the new generation of FS handles graphics in a very different way to its predecessors. By comparison, earlier versions lack sophistication and FS2000 users in particular are familiar with dramatic and unpredictable dips in frame rates. The reason for the improvement is due to improved code in two areas: texture and scenery loading and in "smart" graphics.

Fluid Frame Rates

Dealing with these in turn, you will almost certainly have noticed the long delays while Flight Simulator loads a new flight. The reason for this is that the sim is loading up all the mesh and textures it thinks it might need - before you need them. Previous versions just loaded up whatever was in the direct line of sight and dealt with new areas by loading them as the plane approached.

The "smart" graphics engine in Flight Simulator also prioritises display tasks in order to achieve the most fluid and detailed display it can, which has an unexpected result - frame rates aren't the be-all and end-all of running the sim any more. Experimentation shows that after flight physics, the highest display priority is given to the panel, followed by the visual model and scenery, with AutoGen and reflective textures trailing a poor third. It isn't difficult to check this for yourself. Just start up Flight Simulator and do a circuit of one of the detailed airports in medium cloud with AI at 100%. Depending on the spec of your machine, AutoGen may completely disappear and so may much of the airport.

The "brute force" graphics engine of FS2000 didn't prioritise, which meant that the only defence against frame rate dips was to build in a large safety margin, hence user's obsession with frame rates. By contrast, if you have Flight Simulator, there is much less need to panic if you find you are running with a lower average frame rate. In effect, the new display engine protects you against "slide shows" and it will trade greater fluidity against fast frame rates - if you let it.

This isn't to say that Flight Simulator doesn't ever get it wrong. Strange though it may seem, the one time you are likely to run into trouble is if you run Flight Simulator in spot plane view on a relatively fast machine with target frame rate unlocked. In this configuration, Flight Simulator can generate sudden frame rates peaks which make the plane appear to leap forward. This is the absolute reverse of what you would expect in FS2000 and it gives us a clue about how to configure Flight Simulator for best results.

Target Frame Rates

The key to smooth, fluid displays in Flight Simulator is held by the Target Frame Rate slider, found in the **display\hardware dialog (Figure 1)**. In all probability you may have wondered what this little control does, because the help files say almost nothing about it. Yet the single most important message about configuring

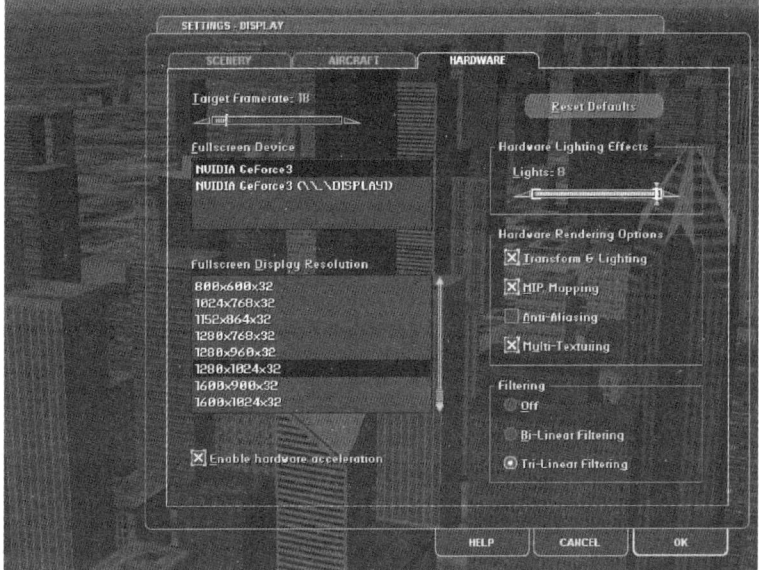

Figure 1

The Good Flight Simmer's Guide Mk.II - 2003   35

Fluid Frame Rates

Flight Simulator is that if you can achieve an average frame rate of 18-20, there no point allowing it rise any higher; in fact, doing so can be counter-productive on anything but the very fastest machines (i.e 2.0 Ghz Pentium IVs or better).

The purpose of the target frame rate control is to prevent the FS engine from allowing the display frame rate to rise any higher than a specified value - and in tests, few people can detect the difference between 18 fps and 60 fps, if the frame rate is stable. What everyone can detect is sudden fluctuations between 10 fps and 40 fps, which we call "stuttering".

The reason for locking frame rates down is best understood through an example. Let's take a 1.7 Ghz Pentium IV with the frame rate locked down to 20. This machine can run at up to 60 fps. If we lock the frame rate to 20, what happens to all the processor power that would have gone into those "extra" 40 fps? The beauty of Flight Simulator is that the display engine doesn't waste them; they are used not only for screen display but to pre-fetch textures and mesh, for AutoGen and to generate lighting, all of which smooth the graphics. On a fast machine, less frames can actually give you a better display.

Running Flight Simulator On Marginal Systems

On a slow machine, adjusting the frame rate down doesn't work as well, for the simple reason that if the system is struggling to maintain even 18 fps, gating the rate brings few benefits. In this case, the only way to improve fluidity is to reduce the load on the CPU which means experimenting with the display sliders, by pulling them back until such time as your PC can regularly produce 15-18 fps.

The nearer a system is to the minimum spec the bigger the trade off between detail and frame rates. At the very lowest level the sim will barely achieve 10 fps and Flight Simulator will look at the way it does when the effects are turned off. Even with a fast machine you might want to read this section because there are circumstances where you may need to protect the frame rate as much as possible - even on a 1.7 GHz+ Pentium IV. The advice which follows will help you to do that.

In each case, the adjustments should be made after selecting "Options" from the main menu, followed by "Settings" and then "Display":

1. Click on the Hardware tab and uncheck anti-aliasing. Speed will increase by about 30-40% extra fps , without any noticeable effect on image quality. If you have an nVidia card, running Flight Simulator in windowed mode does the same thing, due apparently to a bug in Flight Simulator.

2. Still in the hardware section, uncheck multi-texturing - this will cause light tiles to appear in the sea from time to time and you will lose the reflective texturing, but the benefit is another 30-40% more fps.

3. Click on the Scenery tab **(Figure 2)** and pull the "maximum visibility" slider back to 24 miles or so. The benefit varies, but can amount to an extra 25% more fps. As an alternative you can download FSUIPC, a freeware utility which has the option of a visibility "fix" which is graded by altitude. Using FSUIPC provides remarkably realistic visibility conditions in Europe, where haze is much more of a problem than in say the American mid-west. Also Download the Documentation that comes with it.

4. Pull the AutoGen density slider back progressively and check the effect on frame rates. Depending on your system you can add 50% or more to the frame rates by zeroing the slider.

5. Pull the water effects the slider back. The sea will look less interesting but you will gain about 5% more fps.

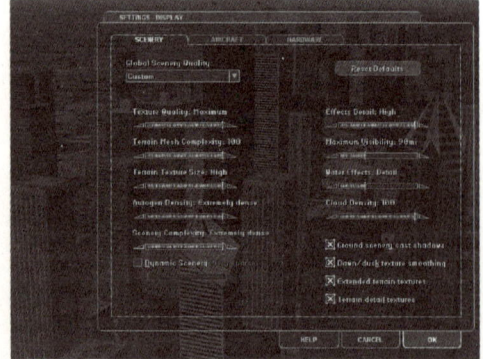

Figure 2

Fluid Frame Rates

6. Turn to Aircraft tab **(Figure 3)** and uncheck Reflections. This will give you another 5% more fps.

7. Pull the AI traffic slider back in the **Options\Settings\ATC** dialog **(Figure 4)**. This can have a dramatic effect near large airports but will have almost no effect in the bush. However, Flight Simulator has an AI resource leak bug. If you ever find your frame rate progressively declining on a long flight, pulling the AI slider back to zero may restore your frame rates.

8. The zero option is always to fly with blue skies. If even this doesn't work, stay well away from detailed airports and limit your flying to areas flat terrain but if you have to do this, you either need a new PC or a copy of FS98.

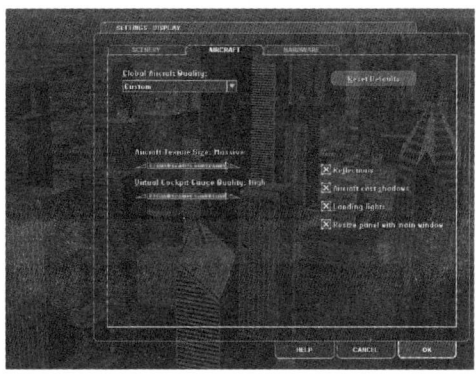

Figure 3

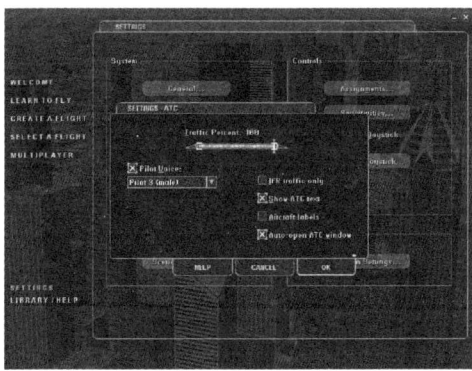

Figure 4

Conclusions

The latest incarnations of Flight Simulator turn the frame rate issue on its head. With previous versions of FS, the only way to be sure of getting a smooth performance when you needed it was to tune the sim to give the fastest possible frame rates - all the time. Flight Simulator's sophisticated new algorithms have turned the frame rate equation upside down, and locking down the target frame rate is THE key to success!

by Andrew Herd
For TecPilot

SECTION three

Learning to Land

Learning to Land

There's an old aviation saying that take off is optional but landing is compulsory. True words indeed, landing is one of the critical phases of flight that must be mastered by all pilots, and it is usually seen as the most significant milestone in early training. Not only should the basic landing be mastered but all the other types of landing need to be mastered too. Safely placing the wheels back on the ground is the second most important skill for a student pilot to learn. The first is, of course, managing to fly around the sky without crashing!

To be able to execute a safe landing in all circumstances, you don't really need to understand what happens. Many pilots fly their whole lives without properly understanding what is going on. However, if you do absorb some of the theory, your chances of landing under those awkward and demanding conditions is dramatically increased.

So let's start with the theoretical side of landing. Don't worry, the theory will not be in too much depth. There will be just enough detail to give you the necessary foundation in the subject.

Angle of Attack Versus Speed

The first piece of theory we need to look at for your landing is how a wing performs when the plane it is attached to slows down.

To think of a wing providing lift above the stall speed, and no lift below the stall speed, would be a gross oversimplification. The lift generated by any wing is dependant on a number of factors but the two most relevant to the pilot are the speed of the air flowing over the wing and the angle of attack. As the speed increases, the lift generated increases. Fly faster and the wing will produce more lift, fly slower and the wing produces less lift. Fly too slowly and the wing will not produce enough lift to support the plane. It is easy to see the effect of speed on lift by flying a plane straight and level and then changing speed without raising or lowering the nose. Fly faster and the plane will climb, fly slower and it will descend.

The angle of attack is a far more complex concept. It is the angle at which the wing passes through the air, which is not necessarily the same as the angle of the plane on the artificial horizon. The following examples should make this phenomenon clearer.

- The wing on a plane that is moving along the runway for take off will have an angle of attack that is around zero. Think about this from the aircraft's point of view, as far it is concerned, all is sees is the air moving past it. The air appears to be flowing over the wing from directly in front to directly behind. On the runway, there is no angle between the direction of the air and a line drawn from the front to the back of the aircraft or wing. This is only a generalisation because the wings are normally bolted onto the fuselage very slightly nose up, and the undercarriage will dictate how the aircraft sits on the runway but we can ignore these complications for this explanation.

Figure 1 – Example of zero angle of attack on the runway

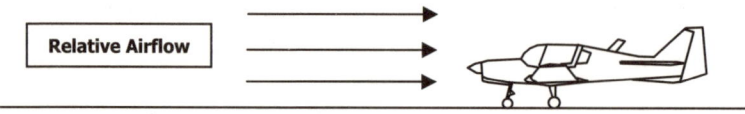

- Next assume that the plane rotates ten degrees nose up for take off but at too low a speed. It has enough speed to rotate but not enough speed to lift off. The plane would be ten degrees nose up but the relative airflow is still directly along the runway. If you think of this from the planes perspective again, the air is moving over the wing from underneath, by an angle that is ten degrees below the nose. This is the angle of attack and it is ten degrees.

Learning to Land

Figure 2 – Example of Ten degree angle of attack on the runway

- Now think of an aircraft that is in high speed level flight, it will be travelling in pretty much the direction it is pointing. The situation regarding the airflow over the wing will be pretty much the same as it is when the plane is on the runway. Think of the plane stationary and the air is moving from directly in front to directly behind. The angle of attack is zero. As before, this is not quite the case and the plane may be slightly nose up and the wings are usually bolted on slightly raised at the leading edge. In level flight, a typical angle of attack over the wings is about four degrees.

Figure 3 – Examples of zero angle of attack in flight

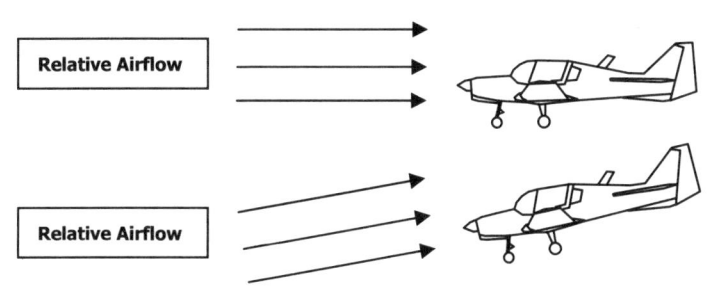

- Finally consider a plane flying quite slowly but with the pilot holding the aircraft exactly horizontal. The plane is descending but the nose is not pointing down at all. In this case the airflow relative to the aircraft is from underneath, just as it was when we rotated the aircraft without taking off. The angle at which the aircraft slides down through the air is the angle of attack. Again this is normally considered from the plane's point of view, whereby the plane is considered stationary and the air passes over it's wings from in front and below, with the angle between a line drawn between the front and the back of the wing, and the airflow being the angle of attack.

Figure 4 – Examples of ten degrees angle of attack in flight

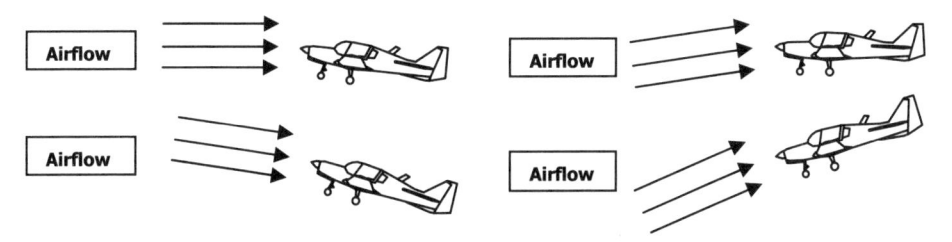

- With the concept of angle of attack firmly in your mind, we now need to think of how it affects the lift that is generated by the wing. As the angle of attack of a wing is increased, the lift increases. A certain angle of attack will be necessary to keep a plane flying straight and level at a fixed speed. If that plane is loaded with more people, cargo and fuel and is to fly at the same speed, it will need more lift to match the additional weight. The only way it can get more lift at the same speed is to raise the nose slightly to increase the angle of attack.

The Good Flight Simmer's Guide Mk.II - 2003

Learning to Land

We are now getting close to what happens during landing. The next step is to think of the interrelationship between speed and angle of attack. As less speed means less lift and a greater angle of attack means more lift, we can trade one against the other. If a plane is slowed down but the pilot does not want to descend, the aircraft nose must be raised to increase the angle of attack to compensate. You can easily try this by setting up a plane in the cruise and then cutting the power. Do this and you will see that to prevent the plane losing altitude you must progressively raise the nose. Eventually, when the angle of attack reaches about sixteen degrees, the wing will give up completely. This is the stall and it can be quite vicious in some aircraft. Remember this trade off between speed and angle of attack, followed by the stall, because it is critical to the landing.

Figure 5 – Trade off between angle of attack and speed for the same lift in level flight

Low speed and high angle of attack.	The lift is always equal to the plane weight in level flight.	High Speed and low angle of attack.

What Is A Good Landing?

The primary objective of landing is to return an aircraft to the ground and bring it to a stop without damaging the plane or the passengers or running off the end of the runway. To achieve this the plane must touch down gently, so that the undercarriage is not pushed up through the wings and the passengers' teeth do not fall out. The plane must also touch down at a relatively low speed; an aircraft touching down at a cruise speed of 300 knots would require an awfully long runway.

Most planes have three wheels or sets of wheels. There are two main undercarriage legs that are mounted close to the wings, which are very sturdy and strong. The third wheel is mounted at either the front or the rear of the plane. Examples of these are the Boeing 737 with the third undercarriage leg at the front, and the Spitfire with the third wheel at the back - the tail wheel. The key fact you take from this is that the main two undercarriage legs are pretty solidly built robust and the remaining one is a relatively feeble affair. Hence your landing technique must subject that third wheel to as little stress as possible. For a nose wheeled aeroplane, this means landing on only the mains undercarriage wheels and then gently lowering the nose wheel onto the ground. For a tail wheeled aircraft, it means either landing on the main wheels first, which is an advanced technique, or a slow "three point" landing.

Speed and Height

One of the strangest relationships when flying aircraft is the relationship between power, pitch, speed and height. The instinctive philosophy is that you should control height with the joystick and speed with the throttle. This works reasonably well in some circumstances but if you try it when flying down the final approach it just does not work. What you need to do is exactly the opposite. You should control your height with the throttle and your speed with the joystick. This all seems the wrong way round and pretty weird stuff until you start to think about it. Actually, most of the technique is really about the order in which you do things and doing them in a way that makes the plane easier to control.

If you are flying down the final approach and are too low, the temptation is to just pull back on the joystick. This has an immediate effect of reducing the rate of descent but there is a rapid secondary effect, which is that the plane slows down. As soon as the plane slows down, as we have found from the theory above, the wings generate less lift. This in turn means that the rate of descent increases again. Hence you are back more or less where you started from, the only difference is that you are flying more slowly. Keep pulling back a bit further on the joystick and you will continue to slow down until eventually you stall. If, instead of pulling

Learning to Land

back, you had added more power, four things happen. First, the propeller blows the air backwards faster and hence it generates more lift from the wings, especially when you have your flaps lowered. Secondly, the plane accelerates a little, and again generates more lift. Thirdly, there is more power available, which is absolutely essential because the shallower your angle of descent, the more power you need to maintain the same speed. Finally, adding some power will tend to pitch the aircraft up slightly. The net effect of all of this is that adding a little power will reduce the rate of descent with very little change in speed. As we have just learnt, any small increase of speed can be easily controlled by raising the nose very slightly. This will slow you down and reduce your rate of descent still further.

In truth, you can make the changes by adjusting either the pitch or the power first. "But", and it is a hugely important "but", your life as a pilot of general aviation aircraft will be very very much easier if you always begin by controlling your height with the throttle and only then follow through with the joystick to hold your desired speed.

Approach to the Basic Landing

The first type of landing to consider is the straightforward normal landing of a nose wheeled aircraft. This is essentially the same whether the plane is a humble Cessna 172 or a Jumbo Jet. Nevertheless, you will find it easier to learn to land in the Cessna, as it is far more responsive and forgiving to any of your mistakes.

The initial phase of the landing is the approach. Never underestimate the importance of getting this part right. It requires a lot of skill and experience to land well from a bad approach and in most cases the pilot is best to "go around" (climb away for another circuit and try again) if the approach looks wrong. What is a good approach? Well you will need the following to be in place to produce a good basic landing.

♦ You should start your final approach far enough away from the runway to give yourself time to do everything you need to do without any rush or panic. A normal approach in a light aircraft would start two miles from the runway and 600 feet above it. However, if you are having difficulty give yourself four or five miles and allow 300 feet height per mile.

♦ You should be flying along the extended centreline of the runway. This means that you are lined up with the runway. When you look out of the windscreen at the runway, it should appear as a straight line running directly away from you, neither inclined to the left nor to the right. The trick to this is to make your heading changes small and to start your correction as soon as you detect any movement away from the centreline. The small heading changes should be five or ten degrees only. It is far easier to stay on the centreline than it is to drift off it and then try to regain it again. Also remember that if there is no wind and you are on the extended centreline of the runway, then if you fly the runway's heading you will not drift far off track.

♦ Your airspeed should be correct and constant. The correct speed is given in the aircraft manual and is often called Vref, however if you do not have these handy, here are some recommendations to get you started.

Sopwith Camel	50 knots
Cessna 172	65 knots
Cessna 182	75 knots
Cessna 208B	80 knots
Extra 300	75 knots
Mooney Bravo	80 knots
Beech Baron 58	95 knots
Beech King Air	105 knots

Remember the theory, you should be controlling this speed by pulling or pushing the joystick very slightly. If the speed drifts a long way from your target Vref, then a change in power may be required, otherwise small changes in pitch are all that is necessary.

Learning to Land

- You should be on the glide slope. Larger airports have lights by the side of the runway called PAPI (Precision Approach Path Indicators) or VASI (Visual Approach Slope Indicators), which indicate whether you are too high or too low on the approach. Generally, half of the lights you see should be white and the other half should be red. If you are too high you will see more white lights and if you are too low you will see more red. The most common is the PAPI, which consist of four lights in a single row at the side of the runway. Your objective is to have two white lights and two red lights. If you are learning to land, make sure you choose an airport with PAPIs. Once you have mastered the art of landing, you will be able to cope quite happily without them. Remember the theory, increase the throttle if you are too low, and reduce the throttle if you are too high. Make sure that the changes you make to the throttle are small. Only when your approach is going very wrong will you need to make big changes to the throttle, and it is normally better to make a small adjustment, wait to see if it is enough, and then make another small change if necessary.

- You should lower the undercarriage and full flaps as early as possible on the approach. You should always lower the flaps one stage at a time and then wait until everything is fully under control before lowering the next stage. As you gain more experience you would leave the last stage of flaps until you are cleared to land and are certain there is no obstructions (such as other planes) on the runway.

- The final 300 feet should be flown at constant speed without having to change anything apart from minor adjustments to your throttle. If you are still trying to master reliable landings, increase this 300 feet to 600 feet or more. By this height, you should have both your landing gear down and your flaps set for landing. This is most important as you need everything smooth and steady if you are to have a successful landing.

These have been given in the order in which you would normally establish them. So the first thing you do is to make sure you have given yourself plenty of space, the next is to establish yourself on the centreline and so on.

Once you are about 100 feet above the runway, you should ignore all your instruments and the PAPI lights. They will only make your life harder than it needs to be. All you need to concentrate on is holding the same attitude and staying on the runway centreline and all you need for this is the view of the runway and the horizon.

The Landing Phase

There are four parts to the landing phase, which are the round out or flare, holding off, touching down and roll out.

Figure 6 - The flare or round out

Having flown down the approach, you should be in a position about twenty to thirty feet above the runway. If you were not to do something soon, you would fly straight into the runway. What you need to do is to convert your descent down the approach into level flight just a few feet above the runway. This is called the flare or round out, and it is achieved by gently pulling back on the joystick. At the same time, smoothly reduce the power to minimum. You should now be flying just above the runway with no power and full flaps.

Learning to Land

Figure 7 - Holding Off

Your objective then becomes to keep the plane flying just above the runway as long as possible, which is why this is called holding off. If you remember the theory, you can trade speed for angle of attack, which is exactly what you must do during the hold off. Each time the plane looks like it is going to sink, pull back on the joystick to stop any loss of height. This should be done in a series of small steps, each time pulling the joystick a little further back and never letting it move forwards again. If you pull back a little to far and balloon (go up again), simply freeze the joystick in position and the plane will eventually lose enough speed to descend again. If your balloon is really bad, apply full power and go around. Be warned that during the hold off, you are likely to lose sight of the runway briefly. This is one of the sins of flight simulators, as in the real thing you would rely on your peripheral vision to confirm that you were still over the runway.

Figure 8 - Ballooning

Once you have raised the nose sufficiently to ensure that the main wheels will hit the ground well before the nose wheel, don't pull the joystick back any further. Now, as we know from our theory, as the speed continues to decay, the lift will reduce and the plane will gently sink onto the main wheels. You have landed. However, do not relax your grip on the joystick, because nose wheels are fragile and still need to be protected. You should either wait for the nose wheel to lower itself onto the runway, which it will do pretty soon, or gently let it down yourself. Having achieved all of this, the roll out should be easy so long as you remember to concentrate.

The Three Point Landing

When an aeroplane has a tail wheel instead of a nose wheel, the landing technique has to be modified slightly. The approach remains essentially the same, although the physics of the landing phase are slightly different. You should round out as before, however the aim in the flare is to hold off until the plane reaches a nose up angle in which all three wheels will touch the ground at the same time. Apart from the need to be more precise about the nose up attitude at touch down, there is no difference between the nose and tail wheel landing....so far. The difficulty is that if you touch down on the main wheels first, the tail descends until the tail wheel touches the ground. However the tail wheel descending further than the main wheels has the secondary effect of increasing the wing's angle of attack. This generates more lift and the plane starts to fly again. This is why so many pilots manage to bounce tail wheeled aircraft on landing.

Figure 9 - Tail wheel bounce

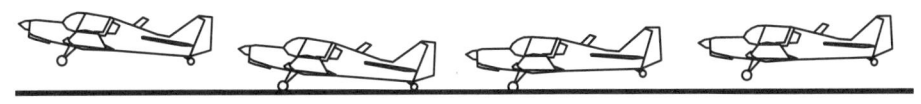

The Good Flight Simmer's Guide Mk.II - 2003

Learning to Land

The second difficulty is controlling the plane in the roll out. Whereas a nose wheeled aircraft always wants to run straight, the venerable tail wheeled aeroplane wants to turn around and point where it has just been. This is the infamous ground loop. The reason is that the centre of gravity is behind the non-steering main wheels (it has to be, otherwise the plane would fall over onto its nose). As soon as the tail gets slightly to one side, the forces want to push it even further out of line. This means that you have to keep ever so vigilant until the plane stops, quickly but carefully, preventing any tendency for the tail to step out of line.

Crosswind Landings

It's a certainty with UK weather that there is always some level of wind. There is also a high probability that the wind is not going to be blowing directly along the runway. Aircraft are rated with a maximum crosswind capability for landings. This is dependant on a large number of factors, but is normally what the test pilot could demonstrate as being safe for that aircraft. However, these limitations will have been determined for the plane by a very skilful pilot and most people use a limit below these. For most light aircraft, the crosswind limit used is about 15 knots, some have limits far in excess of this.

The method used for the approach is exactly the same, with the sole exception that the heading you will need to fly to stay on the runway's extended centreline will not be the same as the runway heading. Approaching the runway in what can seem a dramatically sideways attitude may take a bit of getting used to but it is essentially no more difficult than when there is no wind at all. Although it looks alarming, you still turn left if you are to the right of the centreline and vice versa. The fun part starts with the hold off, as you fly merrily sideways along the runway. If you touch down like this without doing anything, you stand a good chance of damaging the undercarriage and tyres.

Figure 10 - The crabbing approach - the pilot must head slightly into wind to hold the plane on the extended runway centreline

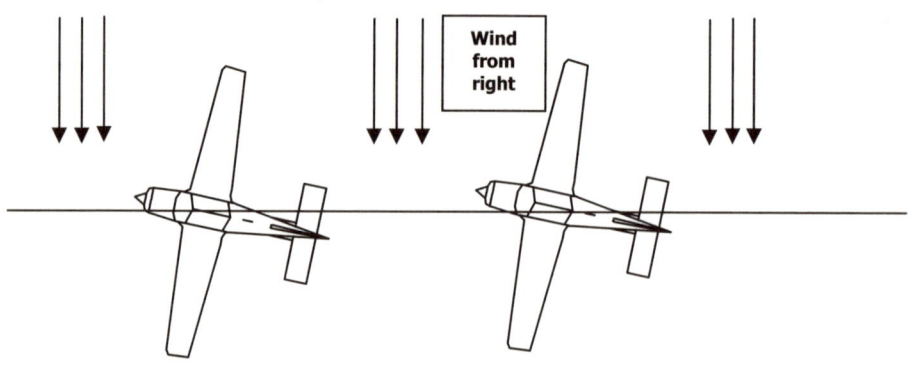

The best technique is as follows. In the final stage of the flare, press the appropriate rudder pedal to yaw the plane so that it is pointing along the runway. At the same time apply a little opposite aileron. So right rudder and left stick, or left rudder and right stick. The amount of stick required is small and should be just enough to hold you on the runway's centreline. You are effectively doing a small sideslip for the last foot or two until one of the main wheels hits the ground. Don't worry, the one on the other side will soon join it. This technique is so effective that it is sometimes easier to land in a strong crosswind than it is to land on a calm still day, when your hold off can take you half way down the runway.

Once you have landed, you will need to keep the stick pushed to the side so that the "into wind" wing does not get picked up by the crosswind. You will also need to apply some rudder to prevent the aircraft weather cocking into wind. If you don't have separate rudder and joystick controls, you may have to modify these instructions to suit your particular set-up. As with all landings, you will need to stay alert until the plane is almost at a standstill. The phrase "it's not over until it's over" is highly applicable to all landings.

Learning to Land

Figure 11 – When to yaw the plane straight for landing

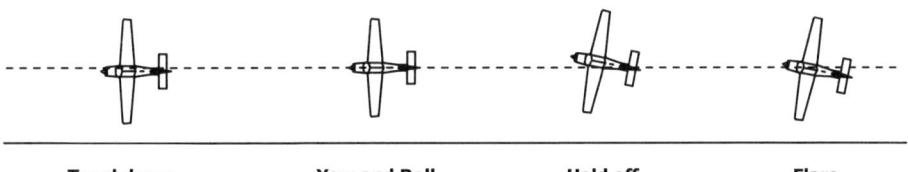

| Touchdown | Yaw and Roll | Hold off | Flare |

Reliable Landings

If you have understood and then practised everything you have just read, you should already be able to execute a safe landing. A trainee pilot will typically fly about 5 hours of circuits before being able to land without the security of an instructor sitting in the left hand seat. Even after the first solo, there will be about another ten hours of circuit flying before true competence is attained. Learning to land can be extremely frustrating for students, especially if the weather keeps them on the ground for a few weekends, during which all the learnt hand eye coordination disappears. Nevertheless, if you keep practising, everything will suddenly click into place, leaving you wondering what was so difficult in the first place. Practise is the key, and pretty soon you will be able to pull off a perfect landing every time.

by Stephen Heyworth
For TecPilot

SECTION four

Guide to Flying Online

Guide to Flying Online

At 8pm (GMT) on a Thursday evening there are over 648 people involved in online flying with various network providers!

Since the early days of Microsoft Flight Simulator people have been attempting to broaden their experiences and one of the main areas of expansion has been flying online with others.

It was not until the release of FS98 that Microsoft introduced "MultiPlayer" into Flight Simulator and "The Zone" as a service on the internet for the application to connect to. The services would allow Flight Simulator to hook up to the Internet and interact with other "players" and exchange information (data), in real-time. The trouble was that the new fangled Multiplayer system was full of bugs, the Internet was not set-up for such huge traffic and basically the system was unpredictable. Further, The Zone couldn't offer a reliable enough environment for reactionary pilots to cooperate with.

Soon after Multiplayer was introduced a group of experienced but disgruntled flight simmers decided to try and overcome the limitations of Multiplayer/The Zone and gathered to develop two key software packages which were later to become known as Squawkbox and Pro-Controller. Bringing up the rear was The Simulated Air Traffic Controllers Organisation or SATCO for short. SATCO oversaw the process of development which seemed to grow at a startling pace. Their aims were simple - to provide virtual pilots (Private and Virtual Airline Captains) a professional minded, reliable backbone to connect to and interact with, on the Internet.

With the launch of FS2000 users had realised that Microsoft had concentrated their efforts on aircraft and scenery design and left the development of Multiplayer for a later release. While FS2000 was technologically advanced, no update was provided for Multiplayer except for a few tweaks to TCPIP. There was still no Air Traffic Control and so users remained frustrated. Pro-Controller and Squawkbox continued to develop and went from strength to strength as more and more Virtual Airlines opened their terminals. Alas, single pilots steered away from Online Activity as it was (and still is) relatively complex to use. This is something we'll be dealing with in this chapter.

It was at this time the FS market saw a sharp upturn in third party freeware and payware development for interactive type products. Add-on adventure packs were released in abundance. They were of exceptional quality for what limited features they offered but lacked functionality. They also required huge resources (e.g. disk, processor and memory capacity) to operate effectively. As individuals continued to steer away from Multiplayer the markets diversified into two distinct groups. One was that of AI (Artificial Intelligence) for single users and MultiPlayer for virtual airline captains and would be Air Traffic Controllers.

The treatment of add-on artificial intelligence traffic finally had its day with the birth of FS Traffic by Lago [www.lagoonline.com] - which was in fact a concept by this books Editor. The application dynamically generated aircraft and ground traffic and was far more complex that Flight Simulator's default dynamic scenery system. This innovative add-on, released during the latter part of FS98's lifetime, gave users a completely new mind boggling perspective to flight simulation. It now placed aircraft movements around airports and planes flying in the skies based on pre-recorded user-defined "tracks". The FS Traffic system gave the same look and feel of an online Multiplayer session without the need to make an internet connection. Further, it gave users the opportunity to recreate accurate airport movements based on real timetables. The tracks could be shared with friends and before long a family of track hungry users developed. Finally, individuals who couldn't get on with MultiPlayer had a friends to fly with - albeit computer generated ones!

Microsoft realised that FS Traffic was a popular add-on with flight simmer's and so decided to include this type of feature, together with ATC, with the release of their eagerly awaited series 8 version of Flight Simulator - FS2002 which was published in the latter part of 2001. This more or less satisfied users who had a hunger for interactivity but it still left a void for the people who couldn't get to grips of connecting to Multiplayer sessions - as it was, and still is, too complex and troublesome to use.

Years on, the fundamentals for creating and logging into a MultiPlayer have not changed since the early days of FS98. The screens and systems used are virtually identical from that of the original. Most individuals are still uneasy about going online, whether they are members of virtual airlines or not. Therefore, the following pages should ease the tension and give you a new perspective about getting online with your mates! It's not as awful as you may think, once you know what software to download and knobs to press.

Guide to Flying Online

The Zone

The Zone is a service provided by Microsoft [www.zone.msn.com]. When visiting the web site users are requested to register with the Game Zone, a process speeded up if you already have ".net passport" set up on your computer. You may have seen ".net passport" when logging into Windows Messenger for example. This system streamlines the process of logging into Microsoft's range of online products and services by using one password only. Once registered you then have the option to download "Zone Server" software, an automatic process, which only takes a few minutes. This configures your computer to allow you to chat to the other users while setting up a session online.

Even though this chapter deals mostly with Microsoft's franchise of simulators, IVAO also includes software for use with Fly and PS1.

Once logged in you need to look under the "SIMS & SPORTS" page and then select "FLIGHT SIMULATOR 2002" for FS games that are available. MSN provide 51+ game rooms under Standard Flights Friendly Skies, and these vary from people advertising Virtual Airlines, to serious pilots who simply want company online. Select a "Game" which looks interesting and this will provide the I.P. address of the person running the session. You will need to make a note of this as you have to enter it into Flight Simulator's MultiPlayer login screen (see below).

The advantage of using the Zone is that it is fairly easy to set up and requires no additional software, other than the MSN CHAT SERVER software, which is totally automatic. On the downside the sessions are limited. In other words you have to fly around the places where users have set up sessions. You can Host your own session of course, using built in multiplayer software but this can lead to a lowering of performance on your own machine and reduced frame rates.

VATSIM & IVAO

VATSIM and IVAO, two providers of online flight simulation, are very similar in structure and both use the same software to connect to the Internet. In fact, they are so similar they even have the same kind of Divisions (Regions) and Countries.

VATSIM is the successor to the original "Simulated Air Traffic Control Organisation" or SATCO. SATCO was originally conceived by Randy Whistler, with the late Ray F Jones second in command. It was one of the first online flying organisations. The "International Virtual Aviation Organisation" or IVAO for short started in 1998 by people who were disillusioned with SATCO and broke away to form their own organisation. Initially they were less structured than SATCO/VATSIM, however as they have grown, both IVAO and VATSIM offer very similar and professionally run ATC and Flying services.

Both organisations require the use of the independently produced "Squawkbox" software package, which is available at the VATSIM and IVAO and the author's web site at [www.simclients.com]. The zip file contains an auto-extract program, which allows a user to set-up the software wherever they want it installed on their hard drive. It will also install a "Common Shape Library" of aircraft into Flight Simulator's Aircraft directory. This "aircraft library" allows a pilot to get a true representation of other aircraft moving around him in MultiPlayer mode, rather than repeated clones of his own aircraft.

A pilot seeing other aircraft as clones of his own is caused by not selecting "Send" & "Receive detailed aircraft information". It's also because of the "Common Shape Library" (CSL) not being installed.

Once you have installed the software you will need to decide which organisation you are going to join. As we have said already, both have different attractions and draw backs and we recommend that you look at what services are offered in the area you intend to fly in.

The Good Flight Simmer's Guide Mk.II - 2003

Guide to Flying Online

VATSIM for example has a very strong UK following and offers a very high standard of Air Traffic Control. Conversely, IVAO has a large following in France and also offers great ATC coverage.

- To join VATSIM go to the main website: **WWW.VATSIM.NET**
- To join IVAO go to the main website: **WWW.IVAO.ORG/HQ**

Once you have registered with your chosen organisation and installed Squawkbox the procedures for using both organisations are very similar. As the Author of this chapter has more experience with logging on and flying with VATSIM we shall use this organisation as an example.

The only difference between logging into VATSIM or IVAO is the server address you enter in the Squawkbox logon screen. More on that shortly...

Squawkbox

This is the heart of online flying with IVAO and VATSIM, as it provides the interface for pilots to speak to each other, or to Air Traffic Control. Time spent getting to know how to use Squawkbox, "SB" as it's lovingly referred to, can save a lot of problems and confusion later. If however, you just want to learn enough to get flying read on...

Create a Flight Sim folder on your desktop and have shortcut's to all your FS programs in it. This saves time when you want to start several programs for a flight sim session.

With the release of FS2002 the original "SB" software was rendered useless as pilots were unable to connect to networks. Drastic measures were needed and so the Software Development Group rushed off and produced a patch (fix), which now allows "SB" to be run outside FS2002. The patch/fix is called SB_HOST, and is included in the Squawkbox download. Unfortunately, to run SB_HOST you must have Flight Simulator running in "windowed" mode (as opposed to full screen) to enable access to the various "SB" menus.

An alternative version is available on the VATSIM-UK website [www.vatsim-uk.net] which allows you to run "Squawkbox" in full screen mode. The VATSIM-UK version comes as a DLL file (Dynamic Link Library). A "Module" which is installed in your modules folder. This, therefore, does not need to be started manually.

Once we have FS up and running and connected to a multiplayer session our next job is to fire up SB_HOST, if using the stand-alone version. Next we need to start up Squawkbox itself. This will run inside the SB_HOST program. We shall then see the start-up screen, as displayed.

In the player name box enter a name you wish for the session, this is a name which is displayed to other aircraft in the multiplayer session. It should be the same call sign as the one you will be using to ATC. This can either be your aircraft registration (G-KEEF or similar). You may want to use your Virtual Airline call sign, BAW793 for example.

As most of us use a TCP/IP connection we now click on "OK" - but only if we are using MultiPlayer. If not then "simply click "No Multiplayer".

Now we move onto the next screen. Click on "Start Search" to begin the login to the multiplayer session. On the next screen, if we leave the address field blank and click on "OK", Squawkbox will then search for any suitable sessions and will find the session you have already set up in Flight Simulator - make sure it is highlighted, and click "Join".

If you selected "No Multiplayer" when you started Squawkbox you will move straight to the next screen without going through the above steps. We now get the Title Screen up with a bar across the bottom titled "Click Here to Get into the Action". Click on this bar and Squawkbox will finally start up.

Guide to Flying Online

Finally, we move to the program itself either running with SB_HOST or as a full screen module if you downloaded the Dynamic Link Library from Vatsim-UK. When you look at the SB_HOST screen you will now see a box at the top with "Welcome to Squawk-box" in green writing on a dark grey background. This text box is very important, as it not only show's what message's you have sent but also what messages ATC has sent out too.

Use the same call sign when flying online. It allows controllers get to know you. You can make good friends this way if you fly regularly."

As this will be your way of communicating with pilots or controllers we shall take some time to look at the various options available to you from this screen.

On the top line "File" and "Help" relate to SB_HOST. They do not play any active roll so for the time being they can be ignored. Next we come to a small button marked "F", this is a short cut button which allows you to open SB's "Flight Management System". This is useful if your Simulator does not have a Flight Management Computer. Along-side "F" there is a second button marked "E". This is used to open the TCAS (Traffic Alert & Collision Avoidance System) and route display box.

We then come to a long thin horizontal box which has blue bars inside it. When connected to the server these bars will show the "LAG" or delay between your PC and the Server, and will change as more people join or leave the session. Below that there is a circle and a button marked either "Ident" or "SBY", this will show if you are using "Mode C" on your virtual transponder. If you are in "SBY" mode the system will only send brief details to the server and this is displayed on the Controllers screen as a "/" symbol. When squawking "Mode C", your full call sign details are sent and the controller can see your aircraft "Blip", which includes your call sign. The "up" and "down" arrows on the next two buttons enable you to scroll messages (UP and DOWN) shown in the grey box below. The next big white box is where you input a message you wish to send. Just click in this box, wait for a cursor to appear and then type your message once you have connected to ATC.

As we have already mentioned the grey box is used to display messages "broadcast" on the frequency you have selected. If however you right click anywhere within this box, you will get the main menu options that you will need to use SB fully. The first thing we must do is to tell SB what call sign we shall be using. Therefore, right click in box and bring up the menu and select "ATC Functions". You will now see a sub-menu with two further options; "File Flight Plan" and "Change Call sign". If we right click on "Change call sign" we get a separate dialogue box with the call sign "NEWBIE" displayed. Overtype this to your chosen one and then hit return to save it.

Before running Squawkbox, examine your Flight Sims Aircraft directory. Make sure that you have a folder called "CSL". If not, re-run your SB installation program to re-install it. It is this folder that lets you see other aircraft correctly.

Next we need to work out a route to fly. In our example we shall depart from Birmingham International, (EGBB). VATSIM try to make flying online as realistic as possible and therefore use "Standard Instrument Departures" (SIDS).

Today we shall use the "Westcott 4D" departure from runway 33. This will route us around other airports and traffic and bring us close to Heathrow's (EGLL) airspace. From the Westcott NDB we can join a "Standard Terminal Arrival Route" (STAR) which will route us to intercept the ILS for runway 27 Right at Heathrow. The route will be "EGBB WCO4D WCO BNN1A EGLL 27R".

We are now ready to enter our route into Squawkbox (SB) ready for when we connect to the VATSIM Server. So, once again, bring up the "Main Menu" and highlight "ATC Functions" but this time right click on "File Flight Plan". This will pop up the Flight Plan dialogue which we need to complete to allow a Controller to know where we are going.

Guide to Flying Online

When first starting off, to avoid embarrassment with heavy landings and "accident's" when flying online, go to the Aircraft Realism menu on Flight Sim and turn off "crash detection". No matter how hard that first bounce, you will still land safely.

Again, starting at the top there are several boxes we need to complete. First we have the Aircraft type selection. As we are using Squawkbox, which has a built in Traffic Collision Avoidance System (TCAS), we click on the TCAS box to show a tick. Today we shall be flying a "PSS 320 Professional" Airbus A319 [**www.phoenix-simulation.co.uk**] so click on the drop down menu and select the "British Airways A319" option. We can ignore the suffix marker for the moment, so it's down to the next block.

We shall be flying Instrument Flight Rules (IFR) so click on that box on the left of the screen.

Next we have the call sign we entered earlier and the aircraft type from the top of the screen. True airspeed is used by other applications that interact with programs from the server. Airspeed is entered in Knots (Nautical Miles). For an Airbus this is typically 280 knots (Kts) and not 120 as displayed here.

The Departure point is the four-letter ICAO code for your departure airport. If you are not sure of the code then the VATSIM-UK web site not only lists the airports by name but also includes all the charts that you will require. Alternatively, don't forget our web site at TecPilot [**www.tecpilot.com**] includes a full SIDS/STARS database and airport listing too.

The next 2 boxes' are for entering the time when you would like to take off. These are entered in Greenwich Mean Time (GMT), Zulu. This is now commonly referred to Coordinated Universal Time or UTC. Although UTC is now the standard, the majority of online flyers and controllers will still refer most time references in "Zulu". Our estimated time of departure will be 14:35 Zulu. As this is only a short flight we shall not be able to climb very high, so we have requested a cruise altitude of 8,000 feet or flight level 080, as this is above the transition level for the area we shall be flying in. 8000ft is also the bottom of the Holding Stack at Bovingdon so we may be able to jump the queue if there is a lot of traffic about.

The large box is where we shall enter the flight plan we have decided to fly. This is entered in "free text" and should include:

- The Departure Airport
- SID (If Applicable)
- Waypoints and Navaids
- Airways
- STAR (If Applicable)
- Destination Airport

All this information is available from charts available from the VATSIM Network or TecPilot and really and add to the realism of online flying.

SID's and STAR's are set up to provide smooth flowing traffic into and out of major airports. On the BNN1A STAR we shall fly will have us leaving Bovingdon at 8000 feet so that traffic leaving Heathrow for the east can climb to 6000 feet, thereby maintaining 2000 feet separation. Once clear of the BPK6G SID route, we shall be cleared to descend for the approach to the runway.

Next we come to the destination box and again we use the four-letter ICAO code for Heathrow (EGLL).

The estimated time of the flight comes next and is entered in hours and minutes.

Remarks gives you a chance to tell the controller that you are a new pilot and a simple remark such as "New Pilot" will enable them to identify that you may need some help with your first flight. Occasionally you will include such remarks as "No Charts" or any special requirements you may have. Don't enter silly remarks like "Any Chance of a Beer" as you will be frowned upon and your session will go extremely slowly or you may be ignored altogether.

Moving down to the bottom block we first enter the estimated Fuel we have on board the aircraft. This can be worked out by dividing the total fuel on board by the average fuel burn per hour for all engines.

Guide to Flying Online

Next is the "Alternate Airport". We use this in case of any problems which could mean we cannot land at our destination. Weather conditions are the most common cause of aircraft having to divert to an alternate airport. We must enter a name in the next two boxes and our base (home) airport in the third. Enter any figure in the last box as it has no bearing in our online flight. That completes this page and we can now click "OK".

This will file the flight plan and this dialogue will disappear. If you look at the grey "Communication" box in Squawkbox, you will now see a new message displayed that says, "Flight Plan Filed".

All this may seem daunting. You will be pleased to know that Squawkbox is actually quite intuitive once you have used it a few times. Most importantly, don't let it overwhelm you. It is surprisingly easy once you know how so keep calm!

Getting Connected

Up to this point we have been setting up "The Session" ready to go online. We are now all set to get connected.

Therefore, right click on the grey box again to bring up "Main Menu" and select "Squawk Standby". This is good practice and ensures that when you connect your blip will not blot (Phew!) out other traffic on the controllers screen. Again, bring up the "Main Menu", but this time select "Connect". This will bring up the "Connect " dialogue which allows us to enter personal details ready to log onto the server.

The first box is the address of the server we wish to connect to. In this case we will enter 212.19.195.26, which is the VATSIM-UK server. If we entered 195.207.29.198, we should be connected to the IVAO server in Brussels. Server addresses are listed on the website of the organisation you are flying with. If there are problems with one server, you can always try logging on to another on the network.

Moving down the screen we now need to enter the Pilot Identity Number and password we have been issued by VATSIM or IVAO. Our name and home base are already entered from the Flight Plan screen.

"Roger Wilco" is a separate utility, which simulates the Radio side of communications and can be started automatically if we enter a server address and tick the box. We will come back to Roger Wilco and voice communications later. Again click on "OK" and we should be connected to the server. Once connected Squawkbox will download the weather conditions for the area we are flying in. These conditions are usually updated every hour.

If you have a microphone, it is easier to "talk" to Controllers using your voice than having to type messages, especially during the approach phase of the flight!

We now need to file our flight plan online so bring up the "Main Menu" and select "Re-file Flight Plan".
Tuning in to ATC

Our communications box should now contain a series of lines (of text) in blue, which remind us of the condition of the server connection. When we connect we will be on the "Unicom" channel which is used whenever there is no ATC cover for the area that we are flying in.

Bring up the "Main Menu" again and this time we shall select "ATC directory". This will bring up a list of ATC stations and the frequencies that they are working on. We'll look for one covering our area and select it. If there are several controllers with an EGBB prefix in their call sign we would connect in following order...

- **Delivery (DEL)**
- **Ground (GND)**
- **Tower (TWR)**
- **Approach (APP)**

... starting at whichever is first in the list. If we have no ATC at the airport we would contact Area Control (CTR) and

Guide to Flying Online

ask if they offer a service for the airport. When we highlight the box and double click on that line we will be connected to that controller's frequency. We'll also receive Automated Terminal Information Service or ATIS.

ALWAYS listen to ATIS Broadcasts as they contain useful information such as runways in use, local QNH and other essential information. If we are using Roger Wilco and the controller has a Voice Channel set up in ATIS we should also be connected to their voice channel too - automatically!
What Should I say?

TIP: "When using voice communication, (Roger Wilco) always think about what you are about to say before saying it! There is nothing worse than getting half way through a sentence and not knowing what to say next. Think "ABC" Accurate - Brief - Clear. "

Going back to our flight, we connect to EGBB_TWR (Birmingham Tower) and listen to what is going on. We will not break into the middle of a conversation as we need to request our clearance. Therefore, bring up the SB screen and click in the white box next to the up and down arrows. Our cursor should now be displayed to the left of the box. Now type a message in the box, in our case:

"Good Evening Speed bird 793 with you stand 61 at EGBB, request IFR clearance to Heathrow - BAW793".

Let's break the message down for you. It doesn't hurt to be polite to somebody who is providing a service to you, hence the "**Good evening**". Next we say who we are and where we are. "**Speed bird**" is the call sign for British Airways, and as **stand 61** is one of the default stands included in FS2002 for Birmingham, we will use that. "**Requesting IFR clearance to Heathrow**", this is what we want to do and what we would like from the controller. We always end with our call sign which, in this case, is "**BAW793**".

The Controller will then respond with a message similar to this: "**Speed bird 793, Cleared to Heathrow FL80, Westcott 4D departure, Initial climb 6000 feet. Squawk 5456**". If you remember, all the information he has read back is in the flight plan we filed earlier. Once we have checked the information provided we repeat the clearance back to the controller and the flight commences in a similar way to that of FS2002's default ATC - but with a REALISTIC SPIN!

As VATSIM and IVAO both provide a full brief for pilots on their website (which includes specific tips for each organisation) we won't undertake a full flight here. Simply go to the relevant Pilot Information section on the web site for more information and tips.

TCAS

We have already mentioned the Traffic Collision Avoidance System which comes as part of the SB package. This screen, when displayed, will show you the position of other aircraft in relation to you. If traffic is within 1000ft and 3 miles it will be displayed as a red diamond, with numbers in the same colour next to it. The numbers relate to the difference in altitude with 0 being traffic at the same altitude. Positive numbers are for traffic above and negative numbers are for traffic below. This combined with range rings and clock face informs you exactly where traffic is. If other aircraft are not in conflict they will be shown in white.

Flight Management System (FMS)

The Flight management system can control your aircraft during a flight but it is not as flexible as some commercial available ones. By entering your route into the FMC (using the keypad on-screen) the route is displayed on the TCAS and can control the autopilot of your aircraft. This is fully explained in the Squawkbox manual. As most people use aircraft which have modern FMC's built in we will skip this section for now.

You can use Squawkbox as a Flight Management Computer (FMC) if your aircraft lacks one. However, most modern commercial aircraft packages include them!

Guide to Flying Online

Virtual Airlines

With the advent of fixed priced Internet Connections and Broadband, we are seeing more and more people flying online - and for longer periods. Virtual airlines have been around for a number of years with people taking flights offline and submitting their logged hours weekly. VA's are usually rank-structured and promotion is based on the number of hours you fly. Over the years this has slowly changed and virtual airlines are now getting more like their real world counterparts.

British Airways Virtual, for example, is probably one of the fastest growing Virtual Airlines at the moment and has over 1000 registered pilots. It offers real-world schedules for pilots to choose from, a full fleet of liveried aircraft and a professional outlook. Further it is probably one of the only Virtual Airlines that actually has signed a letter of agreement between themselves and their real world counterpart allowing the use of the British Airways name and Trade Dress. Visit the website [**www.bavirtual.co.uk**] for more information.

Due to the sheer number of Virtual Airlines (VA's) out there we recommend you visit our website at [**www.tecpilot.com**] and select Virtual Airlines from our Internet Links database. This will provide you with a register of airlines and links. Be aware, VA's have a short shelf life - so be careful when choosing one or it may suddenly "go bust" - in the virtual manner, you understand.

There are now Virtual Military organisations starting up as well. These allow pilots to fly military aircraft online too. These flights vary from Fighter patrols to Virtual Air to Air re-fuelling, with tankers and other service aircraft flying in close formation. There is even a "Virtual Royal Air Force" with a website [**www.vraf.org.uk**].

Look at several Virtual Airlines before joining one. Here are a few helpful pointers:

- **What's its size? (How many pilots, aircraft, affiliations)**
- **What will it offer you as a pilot?**
- **How often do people fly?**
- **How often is the website updated?**
- **How long it has been operating?**

Many Virtual Airlines start up and close down within a matter of months so go for a well-established airline with an active membership package and incentives!

Roger Wilco

This is an independent program manufactured by GameSpy Industries [**http://rogerwilco.gamespy.com**] and supplied as shareware (Try Before You Buy). Roger Wilco (RW) acts as your radio console and allows you to speak to controllers and pilots online. RW has been selected as the voice communications system for VATSIM.

Before we use RW we need to set it up correctly to get the best performance out of it. Start it up and click on the "Adjust" tab, then click "Configure" to set up the program. The fields are all self explanatory.

Pro-Controller

No article about online flying would be complete without a mention of the other side of the service provided by VATSIM and IVAO controllers. The people who provide online ATC use a system called "Pro-Controller". This basically gives them their radar scope's. The system works because Squawkbox transmits your position in latitude and longitude to the server. Pro-Controller then collects the information from the server and displays it on a screen in the correct position in relationship to other aircraft and navigational aids or airports.

Guide to Flying Online

The screen is made up of the main "scope" display: A communications box below and to the left and system message box to the right.

All text transmissions on the controllers "frequency" are displayed in the communications box. Messages like local METAR handoff requests from other controllers who have checked your ATIS is displayed in the system message box.

Function keys allow the controller to turn various Navaids on and off and to zoom in and out. Each controller has a sector file for the position they are working which displays airports, Navaids and airways etc.

Advanced Simulated Radar Client

This is a new development which at the time of writing is still in the testing phase and once fully developed will replace Pro-Controller. It will provide an all-in-one package for voice and radio communications for Controllers and will revolutionise the ATC side of online flying. You can find out about ASRC by visiting their web site: [www.asrc.info]

Pilot Training

Both VATSIM and IVAO mainly deal with the controller training side of virtual aviation. However, there are areas on their web sites designed to guide pilots through the complete online flying experience. VATSIM are looking towards introducing more formalised pilot training sessions in future. British Airways Virtual regularly offer training flights for their pilots and have appointed special training captains for various aircraft types within their fleet.

The Future

With the advent of newer and better software for online flying, coupled with cheaper internet access, the future for online flying is promising. New software is being released all the time. Users can also use programs such as "Whazzup" (Available on the IVAO website) or "ServInfo", which allows users to see where ATC is available and plan their flights accordingly.

By Keith Davies
For TecPilot

FlightSim Central

WWW.FLIGHTSIMCENTRAL.COM

For the latest software and hardware releases take a trip over to FlightSim Central. Your Friendly Online Pilot store!

- We stock all popular and current flight simulation products made by the biggest and best names out there.

- We have the best hardware in stock from top brand names and at great prices! Check with us FIRST!

- We ship anywhere in the world and in record time!

- A simple, secure, encrypted connection takes the stress out of ordering online. Quick, easy and safe!

- We'll impress you with our great range of Videos, DVDs and Real Aviation Merchandise!

Proud to be associated with:
'The Good Flight Simmer's Guide Mk.II - 2003'

Flight Sim Central, LLC
820 Cartwright Road - Suite 101
Reno, NV 89521

Tel Toll-Free: (800) 477-7467
Tel: Direct & Int'l: 001 (775) 847-9400
Fax: (775) 847-4705

SECTION five

Build Aircraft & Scenery Using GMAX

**ADVANCED CHAPTER FOR
TECHNICAL FLIGHT SIMMERS**

Build Aircraft Using GMAX

What Is GMAX?

gMax is a FREE, 3d development software designed specifically for game enthusiasts by Discreet. gMax gives you the tools you need to create your own environments, vehicles and characters to play within your favourite gMax enabled games. You can alter the game experience by changing something as simple as an object's colour, or you can customize the entire look of the game by creating all of your own characters and elements. How much of the game's look you change is up to you and how far you take your imagination.

What Can I Create For Flight Simulator?

gMax gives you the tools to customize two major elements in FS2002/04, aircraft and scenery. Aircraft can be modelled (digitally sculpted) and animated in gMax, exported, and then flown in FS2002/04. You can also add your own colours and logos to your planes, customizing FS2002/04 even more.

You can use gMax to create the airplane you always dreamed of flying, or even create an aircraft that never existed at all. gMax also assists you in creating your own custom scenery for FS2002/04. You can model a small neighbourhood or an entire city, then fly over it in your own custom aircraft. With the combination of gMax and FS2002/04, you have precise control over where your scenery is placed and how it will look inside the game.

How Does It Work?

Below are the major steps to creating aircraft and scenery in gMax and exporting them to FS2002/04.

- **Install gMax.**
- **Install the MakeMDL plugin.**
- **Model your plane in gMax.**
- **Create your planes textures and apply them to your gMax model.**
- **Add animation if desired.**
- **Create a folder structure for your aircraft in FS2002/04.**
- **Export your plane from gMax to FS2002/04.**
- **Model your scenery in gMax.**
- **Create your scenery textures.**
- **Export your scenery from gMax to FS2002/04.**
- **Launch FS2002/04 and fly your new plane over your new scenery.**

Is It hard To Make A Model For Flight Simulator?

How easy your plane is to model depends on how detailed the aircraft is you want to create. One way to determine how difficult an aircraft would be to model is to examine it to see if it is made out of primitive objects such as spheres and boxes. If you stick with an aircraft that is comprised of multiple primitive shapes, you will be able to build your model much faster because there will be much less complex tweaking of the surface. An example of this is an air balloon, which is essentially comprised of a stretched sphere for the balloon and a box for the basket.

In the examples that follow, we shall create an aircraft and objects for use in Microsoft's Flight Simulator 2002. The process is should be exactly the same for Flight Simulator - "A Century Of Flight" (FS2004).

NOTE: We have NOT reproduced gMax screenshots in this section as greyscale (B&W) does not reproduce them accurately. They are best viewed in colour and on screen. The aircraft we are building is the flight model of a Concorde SST. Over 140 images, related captions and backdrops can be freely downloaded from our web site at: www.tecpilot.com. On the right of our home page is a link called "GFSG-2003 Concorde SST". Alternatively, please send an e-mail to: support@tecpilot.com and we shall happily send you a zip file that contains the unallocated (but not essential) reference material used in this section.

Build Aircraft Using GMAX

Installing GMAX

If you installed the complete version of FS2002 you can find the gMax application on your hard drive in the folder **c:/program files/Microsoft games/FS2002/gMax**. If you did not install the complete version of FS2002 you can download gMax from the Discreet website.

To install GMAX:-

If you installed the complete version of FS2002, navigate to: c:/program files/microsoft games/FS2002/gMax
OR
If you downloaded gMax:

1. Unzip the file and navigate to the unzipped folder.
2. Double click on **gmaxsetup.exe** to launch the gMax installer.
3. Click Next.
4. Click I agree to accept the Software License Agreement, then click Next.
5. Select the folder you would like the software to be installed in, then click Next.
6. Click Next again to begin installation.
7. Click OK to install the Active X Controller.
8. Click Finish when the application is complete.

The GMAX Interface

Before going through this article we strongly recommend you go through the gMax tutorial "Getting Started", found in the gMax tutorial files section under the Help menu.

Adding the FS2002 MakeMDL Plug-In for GMAX (If Required)

MakeMDL (MakeMDL.exe) is a program that converts the objects you make in gMax into objects that can be read by FS2002. We shall use the MakeMDL plug-in from within gMax to export .bgl (scenery) and .mdl (aircraft) files.

Installing the MakeMDL Plug-In

1. Download our zip file containing the MakeMDL Plug-in from TecPilot's web site [**www.tecpilot.com**]
2. Unzip the file to your hard drive. For example in C:/
3. Navigate to the unzipped MakeMDL_SDK folder. The default location is: c:/FlightSim_plugins/plugins. Your path may be different depending on your installation.
4. Copy MakeMDL.exe.
5. Paste MakeMDL.exe in the c:/gMax/plugins folder.

What are Polygons?

Polygon meshes are surfaces that give the user a lot of flexibility in modelling because they can be used to produce high-quality, smooth surfaces, as well as low-resolution, fast-rendering surfaces. Games require a small amount of polygons per surface in order for the game engine to render them in real time. As a result, polygons are the choice for game and Internet developers. We shall be using polygons throughout this tutorial to create our airplane and scenery.

Build Aircraft Using GMAX

What are Primitives?

Primitives are the building blocks of 3d modelling. They are pre-made sets of geometry used to simplify the creation of more complex shapes. gMax has nine Standard Primitives: Box, Sphere, Cone, GeoSphere, Cylinder, Tube, Torus, Teapot, and Plane. We will use these primitives throughout this tutorial to give us a head start in modelling the more complex shapes of our plane. Notice how extensively we use primitives; re-using these techniques will speed the creation of your future gMax projects.

TIPS FOR CREATING A BACKGROUND REFERENCE	
Find references (photos or drawings) of the plane taken directly from the side, front and top views. TecPilot has a huge resource of hi resolution images at [**www.tecpilot.com**]	Use the same size file for each of the views. It can help to start with all the views in one file.
Make sure the scale of all the views is the same. The more precise you can be in this stage the easier your plane will be to model.	Use a dark Gray as the background of your reference images.

Adding a Background Reference in GMAX

1. Launch gMax.
2. Click in the Top view to activate it.
3. Choose Views > Viewport Background. [*RESULT: The Viewport Background options open*]
4. Set the Viewport dropdown to Top, then click the Match Bitmap radio button. [*RESULT: Match Bitmap uses the size of your original image file. This helps keep all of the background images the same size within gMax.*]
5. Click Files and browse to the reference image you want to use for the Top view, then select Open. [*RESULT: If you cannot see the image's name, you may need to change the Files of type dropdown to match your images file type (i.e. BMP or JPG).*]
6. In the Viewport Background options click Apply. [*RESULT: The reference image appears in the Top view.*]
7. Change the Viewport dropdown to Front, then click the Match Bitmap radio button.
8. Click Files and browse to the reference image you want to use for the Front view, then select Open.
9. In the Viewport Background options click Apply. [*RESULT: Your reference image appears in the Front view.*]
10. Repeat these steps to add a reference image to the Left view.
11. Click OK to close the Viewport Background options. [*RESULT: You now have an image in the Top, Front, and Left views to use as a reference for modelling your plane.*]
12. Select the Left view, press G to hide the Left views grid.
13. Repeat the last step to hide the grid in the Top and Front views.

Aligning Each Of The Images To Each Other

Now that we have a reference image placed in each view, we need to confirm that they are in alignment with each other. We will create a box object and use it to align all of the background images together. Once aligned, we can hide the box and begin our model.

1. From the Views menu select Viewport Background. [*RESULT: The Viewport Background options open*]
2. Set the Viewport dropdown menu to Top.
3. Turn on Lock Zoom/Pan, and then click Apply. [*RESULT: Lock Zoom/Pan locks the geometry and reference image together. This allows you to zoom and pan the view without disturbing the geometry's relationship to the reference image. This will become clearer as we walk through the following steps*]
4. Repeat this step for the Front and Left views.
5. Click OK to close the window.

Build Aircraft Using GMAX

6. In each view, Zoom out until you can see the entire background image.
7. In the Create panel, select Box under the Object Type rollout.
8. In the Front view, drag from the top left perimeter of the plane to the lower right, and then release the mouse.
9. Without clicking, drag up until the box reaches the tip of the nose; click once to finalize the Box.
10. Select the box using the Move tool.
11. In the Top view, select the Y axis and drag up until the tip of the box is aligned with the tip of the plane's tail.
12. Using the Non-Uniform Scale tool, scale the box along the Y axis until one end of the box reaches the tail and the other end reaches the nose. [*RESULT: Two of the views are correctly aligned, but the Left view is still out of alignment*]
13. Middle Mouse Button (MMB) click in the Left view to select it.
14. Press CTRL+Alt+B to turn off Lock Zoom/Pan in the Left view. [*RESULT: With Lock Zoom/Pan off the reference image and the geometry can be manipulated independently. Since they are now moving independently we can align the reference image with our box*]
15. Zoom in and out until the outside of the plane is aligned with the outside of the box.
16. Press CTRL+Alt+B to turn on Lock Zoom/Pan. [*RESULT: It is good to keep on view lock so you don't throw your image out of alignment from your objects.*]
17. Select the box.
18. In the Display panel click Hide Selected. [*RESULT: The box becomes hidden from view*]

Modelling The Fuselage

1. Select the Front view with the middle mouse button (MMB).
2. In the Create panel, under the Object Type rollout select Cylinder.
3. In the Front view, click on the centre of the nose and drag to the outside of the fuselage, then release the mouse.
4. Drag up until you see 300 in the Height field under the Parameters rollout, then click to finalize the Cylinder. [*RESULT: Creating objects in gMax is a three-step process. First you click and drag to determine the width or diameter of the object. Then you release the mouse and drag to determine the object's depth. When your object is the depth you want it to be, click again to finalize the object*]
5. Click the Cylinder in the Left view using the move tool.
6. Click the X-axis on the Gizmo and drag left until the end of the object lines up with the left side of the tail wing.
7. Select the Non-Uniform Scale tool and scale the Cylinder along the X-axis until it's right end reaches the cockpit window.
8. In the Modify panel, under the Parameters rollout set the Sides to 10 and Cap Segments to 2.
9. RMB click on the Cylinder and choose Convert to > Convert to Editable Poly. [*RESULT: Converting to Editable Poly allows you to edit individual pieces of the Cylinder, such as a Polygon or Vertex.*]
10. RMB click on the Cylinder and choose Sub-objects > Vertex. [*RESULT: The Cylinder's vertices turn on*]
11. Dolly, Zoom, and Track in the Perspective view until you have a clear view of the nose end of the cylinder.
12. Click the middle vertex on the End Segment to select it. [*RESULT: The Vertex turns red to show that it is selected.*]
13. In the Modify panel, under the soft selection rollout, click Use Soft Selection. [*RESULT: The vertices around the centre Vertex turn yellow. By turning on Soft Selection, you can affect more than one vertex at a time.*]
14. Using the move tool in the Left view, drag the gizmo's X-axis to the right until the selected vertex reaches the tip of the plane's nose. [*RESULT: Because Soft Selection is turned on the second row of Vertices also moved but only at a specified percentage of the selected Vertex.*]
15. Select the third row of Vertices by dragging a square around them using the select object tool.
16. Move the row of Vertices to the left until they are aligned directly behind the cockpit window.
17. Scale the row of vertices until the top and bottom is aligned with the outside of the plane reference.
18. Select the second row of vertices and use the move tool to align the row halfway between the tip of the plane and the third row of vertices.
19. Scale the second row of vertices until the top and bottom is aligned with the outside of the plane reference.

Build Aircraft Using GMAX

20. Select the end vertex and move it into alignment with the tip of the plane.
21. Select the rest of the rows one at a time and align them with the outside of the plane reference. Leave the last row near the tail of the plane untouched.
22. In the Perspective view Dolly, Zoom, and Track until you can see the tail end of the Cylinder.
23. Click the middle vertex on the End Segment to select it. [*RESULT: The Vertex turns red to show that it is selected*]
24. In the Modify panel, under the soft selection rollout, click Use Soft Selection.
25. Using the move tool in the Left view, drag the gizmo's X-axis to the left until the selected vertex reaches the tip of the plane's tail.
26. Zoom in and adjust the last two rows of vertices and the end point until they are aligned with the tail of the plane. [*RESULT: You may need to turn off Use Soft Selection to keep the rows straight*]
27. RMB click on the Perspective title in the Perspective view and choose Smooth + Highlights. [*RESULT: Smooth and Highlights switches the viewing mode from wire frame to shaded mode*]
28. RMB click on the Perspective title in the Perspective view and choose Edged Faces. [*RESULT: Edged Faces displays the wire frame of the object on top of the shaded surface*]
29. The Fuselage is complete!

Modelling The Wingspan

1. In the Create panel click Box.
2. In the Top view drag from the upper Left corner of the wing to the lower right corner of the wing, then release the mouse.
3. Drag down until the Height reads -6, then click to finalize the Box.
4. In the Create panel, under the Parameters rollout change the Height Segments to 2, Width Segments to 3, and the Length Segments to 6.
5. In the Modify panel, choose Taper from the Modifier List dropdown.
6. Under the Parameters rollout, set Primary to Y and Taper Amount to 1.3.
7. In the Modify panel choose XForm from the Modifier List dropdown. [*RESULT: Adding the Xform Modifier allows everything in the stack to be transformed, in this case the taper and the Box, not just one or the other.*]
8. Non-Uniform Scale the Box until it is the same width as the wingspan.
9. RMB click on the Box and choose Convert To: > Convert to Editable Poly. [*RESULT: This collapses the stack, the Xform and Taper modifiers, and allows us to select individual pieces of the object.*]
10. RMB click on the box and choose Sub-object > Vertex. [*RESULT: The Box's vertices appear.*]
11. Select and Non-Uniform Scale each horizontal row of Vertices to match the shape the plane's wing.

Modelling The Engines

1. In the Create panel click Box.
2. In the Front view drag a square from the upper left, to the lower right of one of the engines, then release the mouse.
3. Drag up until the Height reads 125, then click to finalize the Box.
4. Set the Length Segs to 1, the Width Segs to 1, and the Height Segs to 1.
5. From the Side view, move the box into alignment with the engine.
6. In the Perspective view Dolly, Track, and Zoom until you can see the Front of the engine Box.
7. RMB click on the Box and select Convert To > Convert to Editable Poly.
8. RMB click on the Box and select Sub-object > Polygon.
9. Click the polygon on the front of the engine Box to select it.
10. Under the Edit Geometry Rollout click Bevel.
11. Click once on the Extrusion's down arrow, then set the Outline to -2. [*The end is slightly extruded first, and then an edge is created from the Outline value change*]
12. Set Extrusion to -10. [*RESULT: -10 extrudes the Front polygon back to carve the shape out of the Box*]
13. In the Side view RMB click on the box and select Sub-objects > Vertex.
14. Drag a square around the three lower right vertices to select them.
15. Move the vertices back along the X-axis until the angle of the Box matches the angle of the background image's engines.

Build Aircraft Using GMAX

16. Click Editable Poly in the stack to turn off Sub-object selection.
17. Select the Box in the Front view..
18. Pressing Shift, drag the box to the position of the next engine. [*RESULT: The Clone Option box opens.*]
19. Set Object to Copy and click OK. [*TIP: Holding Shift when dragging objects will make a copy of them. You can create multiple copies an equal distance away from the original object by changing the Number of Copies value in the Clone Options.*]
20. With the new engine selected, CTRL click on the original engine to select both of the engines.
21. Pressing Shift, drag the two engines to the other side of the plane, release the mouse when the engines are in alignment with the opposite two engines.
22. Click OK when the Clone Options open. [*RESULT: A copy of the original engine is made*]

Modelling The Tail Wing

1. In the Create panel click Box.
2. In the Left view, click and drag from the upper Left corner, to the lower right corner of the tail wing.
3. Release the mouse, then drag up until the Box's Height equals 5, then click to finalize the box.
4. Under the Parameters rollout in the Create panel, change the Height Segments to 3, Width Segments to 4, and the Length Segments to 4.
5. RMB click on the Box and select Convert To > Convert to Editable Poly.
6. RMB click on the Box and select Sub-objects > Vertex.
7. Drag a square around the leftmost row of vertices, using the rotate tool around the Z-axis; rotate the row to match the angle of the tail wing in the reference image.
8. Move the row into alignment with the edge of the tail.
9. Move the second row of vertices into alignment with the top edge of the tail.
10. Non-Uniform Scale the last three rows down until they are aligned with the tail shape.
11. Select all of the objects in your scene by dragging a square around them using the Select Objects tool.
12. Click on the colour swatch next to Multiple Selected in the Create panel to open the colour picker.
13. Select a basic colour for your plane.
14. Click OK to close the window. [*RESULT: Your model is now all the same colour*]

Naming Your GMAX Objects

As you create each new object in gMax you should give it a recognizable name. Select the plane's fuselage object and at the top of the modify panel name it nose, for nose of the plane. Select the plane's tail wing object and name it tail, for the rear wing of the plane. Select the plane's main wingspan object and name it l_wing. These names coincide with the naming scheme FS2002 uses to identify specific parts of the aircraft. For more information on naming objects in FS2002 refer to the chart file named MakeMDL.doc.

Changing Units In GMAX...

In the plane tutorial we used Generic Units but you can change the units to read as feet and inches or metric units. This option is useful if you know the specific measurements of your aircraft. To change the Unit Type in gMax:

1. From the Customize menu select Units Setup. [*RESULT: The Units Setup options open*]
2. Select Metric or US Units as the Unit Type.
3. Click OK to close the options.

The Good Flight Simmer's Guide Mk.II - 2003

Build Aircraft Using GMAX

GENERAL MODELLING TIPS	
We have built our model with slightly square edges. We added extra geometry (lines) to our edge surface to later add smoothness to the edges.	You can add more detail to the backside of the engines by repeating the Bevel steps used to detail the front side of the engines.
You can also experiment with selecting the edges of the wings polygons and moving them out away from the main wing. This will make the edges more sharp and rounded looking.	One thing to keep in mind is that much of the fine detail can be done during the texture stage. You will get better FS2002 performance by having fewer polygons on your model and a more detailed texture.
To smooth the edges, experiment with scaling the vertices along the edges.	

Exporting Your GMAX Model To Flight Simulator (e.g. FS2002/2004)

There are a few essential steps to setting up your models to be read by FS2002. We shall walk you through each one.

Creating A Folder Structure For Aircraft

1. Browse to; **c:/program files/Microsoft games/FS2002/aircraft** to view the aircraft folder structures already created for Flight Simulator. As you browse around notice that each aircraft has these common folders and files:

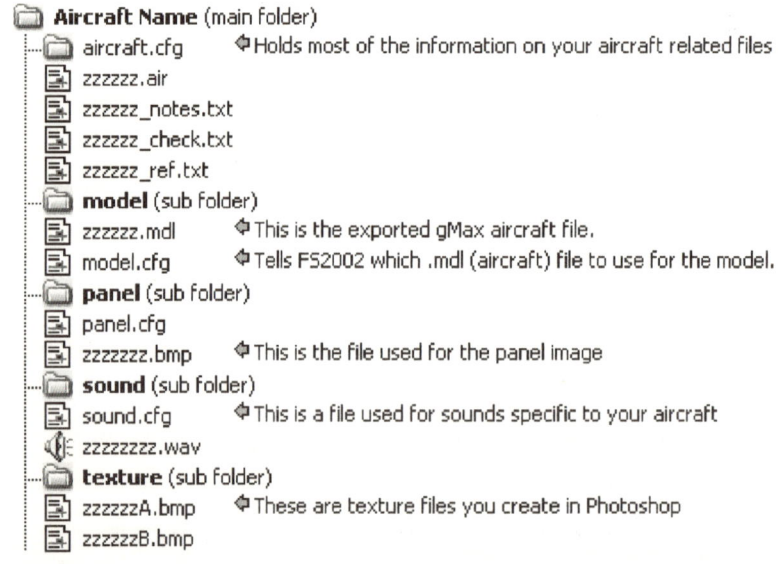

   - **Aircraft Name** (main folder)
     - aircraft.cfg ⬅ Holds most of the information on your aircraft related files
     - zzzzzz.air
     - zzzzzz_notes.txt
     - zzzzzz_check.txt
     - zzzzzz_ref.txt
     - **model** (sub folder)
       - zzzzzz.mdl ⬅ This is the exported gMax aircraft file.
       - model.cfg ⬅ Tells FS2002 which .mdl (aircraft) file to use for the model.
     - **panel** (sub folder)
       - panel.cfg
       - zzzzzzz.bmp ⬅ This is the file used for the panel image
     - **sound** (sub folder)
       - sound.cfg ⬅ This is a file used for sounds specific to your aircraft
       - zzzzzzzz.wav
     - **texture** (sub folder)
       - zzzzzzA.bmp ⬅ These are texture files you create in Photoshop
       - zzzzzzB.bmp

2. Now that we have a basic idea of how the folder structure should be set up. Let us use this knowledge to make our own.
3. Inside the main FS2002 aircraft folder, create a new folder and name it carbon100. [*TIP: You can name your folder whatever you want. We shall use carbon100 for consistency throughout this tutorial*]
4. From the b737_400 aircraft folder, copy aircraft.cfg, Boeing737-400.air, Boeing737-400_check, Boeing737-400_notes, and Boeing737-400_ref and paste them into the carbon100 folder. [*TIP: The Boeing737-400 has the fastest airspeed so it is the closest match to our concord. Whenever you create a new aircraft, use the files of the plane whose attributes are closest to your plane's attributes*]
5. Change the Boeing737-400_check filename to: **carbon100_check**.

Build Aircraft Using GMAX

6. Change the Boeing737-400_ref filename to: **carbon100_ref**.
7. Change the Boeing737-400_notes filename to: **carbon100_notes**.
8. Change the Boeing737-400.air filename to: **carbon100.air**.
9. Open the aircraft.cfg file in the text editor Notepad.
10. We shall be working with the area under [fltsim.0] in the aircraft.cfg file.
11. Change the title from Boeing 737-400 to **Carbon 100**.
12. Change the sim from Boeing737-400 to **Carbon100**.
13. Change the ui_manufacturer from Boeing to **Carbon**.
14. Change the ui_type from 737-400 to **Series 100**. [*TIP: The ui_type is the aircraft name which appears in FS2002 as the Aircraft model*].
15. Change ui_variation from American Pacific Airways to **Carbon College Airways**. [*HINT: The ui variation appears in FS2002 as the name of your colour varient or "livery"*]
16. Change the current description to: "The Carbon 100 is the best aircraft ever created." [*HINT: The description appears in the airplane description area of FS2002*]
17. Change the kb_checklists from Boeing737-400_check to **carbon100_check**.
18. Change the kb_reference from Boeing737-400_ref to **carbon100_ref**.
19. Save and close the aircraft file.
20. In the carbon100 folder create new folders named model, panel, sound, and texture.
21. Copy model.cfg from b737_400's model folder and paste it into carbon100's model folder.
22. Open the model.cfg file in the text editor and change B737_400 to **carbon100**.
23. Save the file.

Creating Your .mdl File In GMAX

1. Open your final plane file in gMax.
2. In the Display panel select Unhide All. [*RESULT: The original reference Box appears*]
3. Select the reference Box and Delete it.
4. From the File menu select Export. [*RESULT: The export options open*]
5. For the Save In folder select: c:/program files/Microsoft Games/FS2002/aircraft/carbon100/model
6. Select Flightsim MDL File (*.mdl) for the Save As Type.
7. Name the file carbon100 then click Save to export the file.
8. Start FS2002 and click change under the Current Aircraft text block.
9. Select Carbon as the Aircraft Model. [*RESULT: Your gMax model appears on the screen in your chosen colour.*]

Adding Animation To Your Models

You can add pre-made animations to your FS2002 plane by using the Stock Animation Names from the makeMDL.doc file. Any gMax object with a part name from the Stock Animation Names table will automatically use its corresponding stock animation, no extra animation in gMax necessary.

Re-Using Old Models to Create New Ones

Now that you have a complete plane model you can use this as a base object for future plane models. Open the plane model in gMax and Non-Uniform scale the entire object nose to tail until you have a plane that is half the size of the original. You can already see how easy it is to do just a few tweaks to the surface and have a brand new plane object moments later. Use the tools learned in the first sections to alter the plane into an original aircraft of your design. Make sure to save your file as a different file name so you don't overwrite the original.

Build Aircraft Using GMAX

Texturing Introduction

You can add realism or make your aircraft unique by adding textures to your gMax models. To produce a texture for FS2002 you need an image-editing program similar to Adobe Photoshop, Photoshop Elements, or Paint, which comes with Windows. We shall use Adobe Photoshop for our examples because it is the most versatile of the three. We shall draw our textures over a wireframe screenshot of our gMax objects and then save the Photoshop file as a Bitmap to be read by FS2002. Once we have finished our textures in Photoshop we will apply each texture to the appropriate gMax material channel. Then we will apply the material to our airplane and export the gMax aircraft to FS2002.

gMax has five materials that can be used on your models inside FS2002. The FS2002 materials include a base texture, damage texture, night map, light map and reflection map. Although the materials do not have the same name in gMax as they do in FS2002, each of these materials can be applied to an object within gMax by following the simple comparison chart below. The FS2002 names are on the Left and the applicable gMax material names are on the right.

FS2002	gMax Material Names
Base	Diffuse
Damage	Used by FS2002 only. No need to import into gMax
NightMap	Ambient
LightMap	Self-illumination
ReflectionMap	Specular colour

Each of these material types has a distinct purpose when brought into FS2002. A texture map applied to the Diffuse channel in a gMax material will determine the look of the main plane object in gMax, known in FS2002 as the base texture. This base texture will include your basic colouring, door and the metal sheet textures of your plane. You can think of the base texture as the colour normally seen on your plane during the day.

The **Damage** Texture is a copy of your **Base** Texture manipulated to add a damaged look to your aircraft. The Base Texture is swapped for the Damage Texture when your plane collides with objects or the ground. Damage Textures are used for aircraft objects only, not scenery objects.

The **NightMap** texture determines what your plane will look like at night. FS2002 blends the base texture with your NightMap texture during dusk or dawn. To have your NightMap material appear correctly in FS2002 you must add your texture to the Ambient channel of your gMax material. NightMaps are more commonly used for scenery objects than aircraft objects.

The **LightMap** is used to illuminate the base texture. FS2002 combines the base texture with your LightMap. To have your LightMap material appear correctly in FS2002 you must add your texture to the Self-illumination channel of your gMax material.

ReflectionMaps are used in games to fake a reflection of the environment on your objects. Games are rendered in real time, which restricts them from having the time to calculate actual environmental reflections. To add a bit of realism you can add a ReflectionMap to your object through the Specular Colour channel in your gMax texture.

Now lets go step by step in the texture creation process...

Build Aircraft Using GMAX

Creating The Base Texture In Photoshop (Adobe)

Creating a visual reference

1. Open Photoshop.
2. From the File menu choose New.
3. In the New dialog box change the Width and Height to 512 pixels and the resolution to 72.
4. Click OK to create the new canvas. [*TIP: Although we are using a 512 by 512 canvas, it is more efficient to create your textures at 256 by 256. Also, If FS2002 performance is slow, resize your textures to 256 by 256.*]
5. Go to gMax and open your plane file.
6. From the Views menu select Viewport Background.
7. In the Apply Source and Display To area, change the Viewport to Front.
8. Under Current, Click Display Background to turn off the display of your background.
9. Press Apply.
10. Repeat the same steps for the Top and Left views. [*RESULT: All of the background images should now be hidden.*]
11. Press OK to close the Viewport Background dialog box.
12. Click the Zoom Extents button.
13. Press Print Screen to take a screenshot of your gMax interface.
14. Go back to Photoshop and choose Edit > Paste. [*RESULT: Your gMax screenshot is now pasted over your canvas*]
15. Zoom out of your canvas by pressing CTRL - several times.
16. Maximize your Photoshop file so it fills the screen.
17. From the Edit menu select Transform > Scale.
18. With the Square Marquee tool, drag a square around the wingspan in the Top view of the model.
19. Select Edit > Copy.
20. Select Edit > Paste to paste the image of the model to a new layer.
21. Temporarily hide Layer 2 by clicking on the eye next to the layer name in the Layers panel.
22. Make Layer 1 active by clicking on the layer name.
23. Using the Move tool, drag the image up until the Left view of the model is cantered in the window.
24. With the Square Marquee tool, drag a square around the Left view of the model.
25. Select Edit > Copy to copy the image.
26. Select Edit > Paste to paste the image to a new layer.
27. In the Layers panel, drag Layer 1 to the trash to delete it.
28. Make Layer 3 active by clicking on the layer name.
29. Link Layer 3 to Layer 2 by clicking in the space to the right of the eye on Layer 4. [*RESULT: A small chain appears to show they are linked.*]
30. From the Edit menu select Transform > Scale.
31. Pressing Shift, scale the layers up until the tip and tail of the plane touch the sides of the canvas. [*RESULT: Pressing Shift scales the selected objects proportionately*]
32. Press Enter to finalize the scale.
33. Click the link on Layer 4 to unlink it from Layer 3.
34. With the Move tool selected, move layer 3 down to separate the two images.
35. Save your file as concordbase.psd. [*HINT: Save the file as a Photoshop file, this will preserve your layers*]

Creating The Plane's Stripe And Airline Name

1. Open the concordbase.psd file.
2. Click the Create a New Layer button in the Layers palette. [*RESULT: A new layer is created.*]
3. Select Layer 3 by clicking on the layer name.
4. Select the Magic Wand and set it's Tolerance to 1, then check "*Contiguous*" to turn it on.
5. Click once on the outside gray around the Left view of the plane. [*RESULT: The outside gray is selected while the plane is not selected. If the entire gray around your plane is not selected press Shift and click on the gray below the plane to add to the selection.*]

Build Aircraft Using GMAX

6. From the Select menu choose Inverse. [*RESULT: The plane is now selected. We have also selected the empty (transparent) area of the layer, which we no not need*]
7. Pressing Alt, click once in the empty (completely transparent) area of the layer to deselect it. [*RESULT: Now only the plane is selected*]
8. Click Layer 3 to select it.
9. From the Edit menu choose Fill. [*RESULT: The Fill options open*]
10. Under Contents set Use to Background Colour.

Adding Windows, Doors and Metal Panels

Creating Windows

1. Create a new layer in Photoshop. [*TIP: If the layer is not at the top of the layer stack, click and drag the new layer above the other layers*]
2. Select the Rectangle tool.
3. Set your options to Make Filled Region and Rounded Rectangle Tool, then set the radius to 2px.
4. Press D to set the colour swatches to black as the foreground colour and white as the background colour.
5. Drag a small square the size of one of the plane's windows.
6. Pressing CTRL+Alt+Shift, drag the window to the position of the next window, then release the mouse. [*HINT: Pressing CTRL+Alt+Shift makes a copy of the original window*]
7. Repeat this step until all of your windows are created.
8. Use the Lasso tool on a new layer to draw the cockpit window as shown, then fill the selection with dark gray.
9. The windows for the plane are now created.

Creating Doors

1. Create a new layer.
2. Select the Rectangle tool; make sure your options are still set to Make Filled Region and Rounded Rectangle.
3. Drag a small square the size of the door. [*HINT: The door is currently filled black; next we will convert the fill to an outline*]
4. Pressing CTRL, click on the door layer to select the contents of the layer.
5. Press Delete. [*HINT: The door is deleted but the outside selection is still active*]
6. From the Edit menu select Stroke.
7. Click OK to finalize the stroke.
8. In the Layers panel change the layer's opacity to 40%. [*RESULT: The door outline is lightened*]
9. With the Rectangle tool still selected drag a small square for the door handle.
10. The door is now complete.

Creating Metal Panels

1. Select the Line tool. [*HINT: The Line tool is hidden under the Rectangle tool*]
2. Create a new layer and drag it to the top of the layer stack.
3. Pressing Shift, draw a vertical line across the plane everywhere you would like a panel. [*HINT: The lines should go outside the outline of the plane to wrap correctly. The lines will be lightened later*]
4. Draw random horizontal lines between the vertical lines to create complete panels.
5. In the Layers panel change the layer's opacity to 5%. [*RESULT: The lines are lightened*]

Colouring The Tail Wing

1. Create a new layer and drag it to the top of the layer stack.
2. Select the Rectangle Marquee tool.
3. Draw a square that overlaps the entire tail wing.
4. Click on the Foreground Colour swatch and select a colour for your tail.
5. From the Edit menu select Fill.

Build Aircraft Using GMAX

6. Click OK to fill the selection with the Foreground Colour.
7. Select the Text tool and set it's attributes to the desired font, text size, and colour for the tail wing letter.
8. Click near the centre of the wing.
9. Type the letter you would like displayed on the tail. I have used C for this example.
10. Select the Move tool and move the letter into alignment with the centre of the wing.
11. Select the coloured box layer.
12. Select the Lasso tool.
13. Pressing Alt click *the points in Figure 77, then release the Alt key to complete the selection.*
14. *Press Delete. [RESULT: The extra colour area is deleted]*

Colouring The Fuselage

1. Hide layer three and five by clicking on the eye next to the layer's name.
2. Select the background layer and Fill it with the desired base colour of the plane.

Painting The Wingspan And Engines

1. Using the concordbase.psd file, apply the techniques you learned in Creating a visual reference and Creating metal panels to create a wingspan and engine colour scheme.
2. Save the final file as concordbase.psd. [*HINT: Saving the file in PSD format preserves the layers for later revisions*]
3. From the File menu select Save as.
4. Name this file concordbase_T.bmp, and set the Format to BMP. [*RESULT: The file is saved and the layers flattened. This is the file we will use for our gMax texture*]
5. Click Save to save the file.

Applying Textures To A GMAX Model

1. Open your plane file in gMax.
2. Select the Fuselage object.
3. In the Modify panel under the Modifier List dropdown, select UVW Map. [*TIP: UV Coordinates are applied to the surface. UV Coordinates are used to place and move the texture on the surface.*]
4. Click the gMax Material Editor button [*RESULT: The Material Editor open*]
5. Click new to create a new material.
6. Name your new material Base Texture.
7. In the Diffuse Map section click the Assign Texture button .
8. Browse to the folder with your BMP texture file in it. Double click the concordbase.bmp texture to open it.
9. Click the apply button to apply the new texture to the Fuselage.
10. The square in the center of the gizmo tells us which direction the texture is being applied. The texture is currently being applied from the front but we want our texture to be applied from the side.
11. Select the Rotate tool.
12. Rotate the texture square around the Z axis until it reads 90. [*RESULT: Because the texture is not the exact size of the fuselage it does not correctly line up with the surface. We shall correct this problem in the next few steps.*]
13. In the Modify panel under the Modifier List dropdown, select Unwrap UVW. [*RESULT: You should now see three titles in the Modifier Stack. Unwrap UVW on the top, UVW Mapping in the middle, and Editable Poly on the bottom.*]
14. In the Parameters rollout click Edit. [*RESULT: The Edit UVW options open*]
15. Zoom out until you can see the entire fuselage wire frame.
16. With the Move tool, drag a square around the entire wire frame to select the vertices. [*RESULT: All the vertices turn red*]
17. Scale the fuselage wireframe until it's width is the same size as the texture image's fuselage width.
18. Non-Uniform Scale the wire frame vertically until the proportions of the Fuselage match the texture image. [*RESULT: If your wire frame is reversed press the RMB and select Flip Horizontal. The wire frame flips horizontally*]

Build Aircraft Using GMAX

19. Close the window.
20. Select the tail wing object and repeat the above steps to apply the texture to the tail.
21. Select the wingspan object and repeat the above steps to apply the texture to the wingspan.
22. Select the 4 engines and apply a basic red texture to them.

Exporting The GMAX File

1. From the File menu select Export. [*RESULT: The Export options open*]
2. For the Save in folder select: c:/program files/Microsoft games/FS2002/aircraft/carbon100/model
3. For the Save as type select Flightsim MDL File (*.mdl).
4. Name the file carbon100 and click Save to export the file. [*RESULT: Click OK to save over the old exported file.*]

Finalizing Textures With The Image Tool

1. Start the ImageTool program (exe). The one you unzipped in the c:/FlightSim_plugins folder earler.
2. From the File menu select Open.
3. Open the concordbase_T.bmp file.
4. From the Image menu select Format > DXT 1.
5. From the File menu select Save As.
6. Browse to this folder: c:/program files/Microsoft Games/FS2002/aircraft/carbon100/texture
7. Name the file concordbase_t.bmp.
8. Click Save.
9. Launch FS2002. The plane's textures are now visible.

Creating New Texture Types Using The Base Texture

Now that we have a base texture established for the plane we can use it to create a variety of other texture types.

To create a Damage Texture you can save the concordbase file as concordbase_D in Photoshop, then use the Burn tool on the front and bottom of the plane texture to imply a charred plane after a crash. You can also use the airbrush tool to paint on earth tones to add dirt colour from the ground.

Since ReflectionMaps are "fake" reflections of real environments, often a photo of a real environment will make a good reflection map. You can also use a chrome photo to gain a shiny effect on your surface.

Part of the aircraft designed in gMax

Build Aircraft Using GMAX

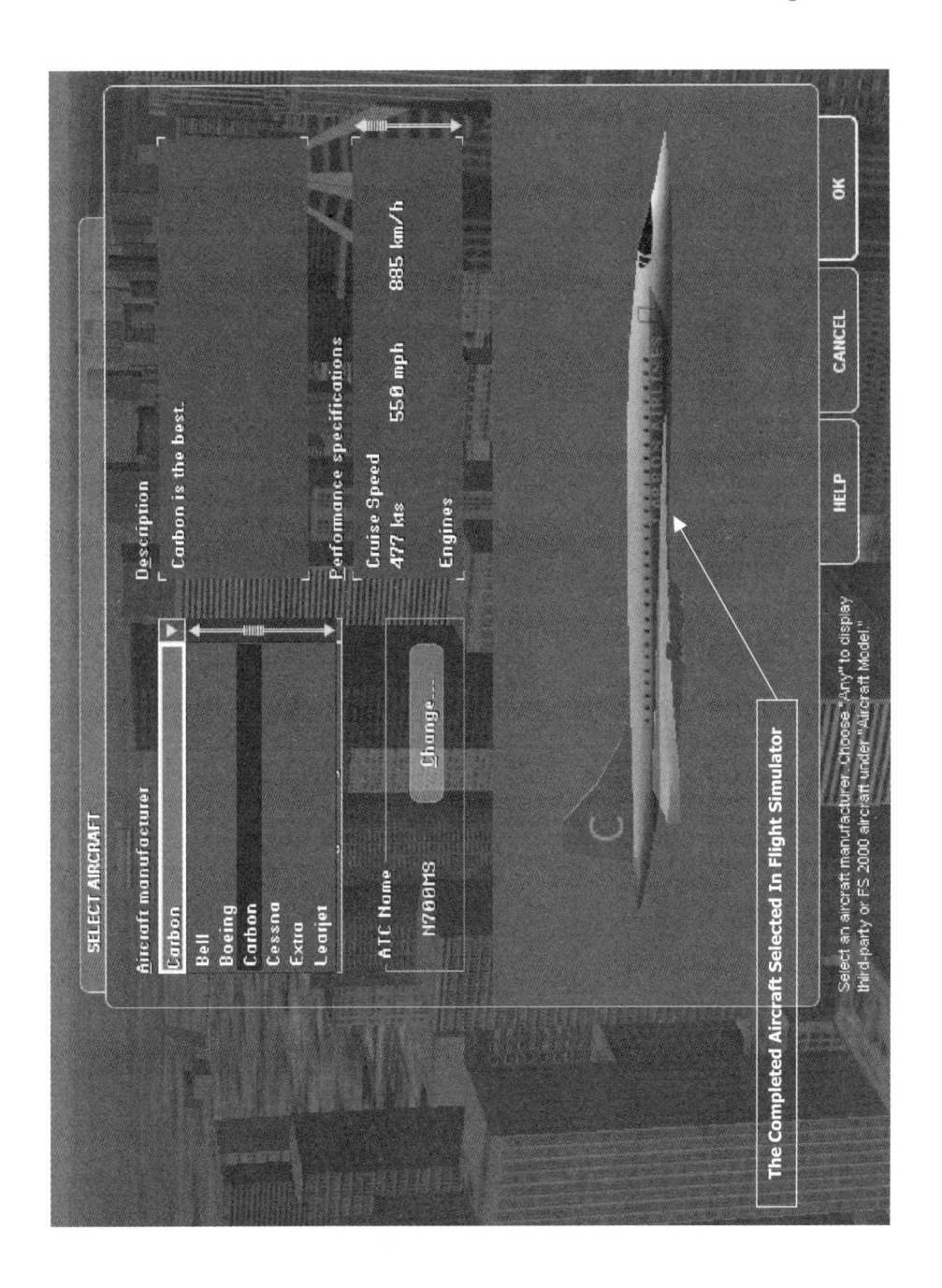

SELECT AIRCRAFT

Aircraft manufacturer
- Carbon
- Bell
- Boeing
- Carbon
- Cessna
- Extra
- Learjet

Description
Carbon is the best.

Performance specifications
Cruise Speed
477 kts   550 mph   885 km/h

Engines

ATC Name
N700MS

Change

OK
CANCEL
HELP

Select an aircraft manufacturer. Choose "Any" to display third-party or FS 2000 aircraft under "Aircraft Model."

The Completed Aircraft Selected In Flight Simulator

The Good Flight Simmer's Guide Mk.II - 2003 — 75

Build Scenery Using GMAX

Creating Basic Scenery in GMAX For Flight Simulator (e.g. FS2002)

Modelling A Simple House Structure

1. In the Create Panel click Box.
2. In the Top view drag a square 150 units in width and 75 units in length, then release the mouse.
3. Drag up until the Height reads 50, then click to finalize the Box.
4. RMB click on the Box and select Convert To > Convert to Editable Poly.
5. RMB click on the Box and choose Sub-objects > Polygon.
6. In the Front view select the polygon facing you.
7. In the Modify menu under the Edit Geometry rollout click Slice Plane.
8. Rotate the slice plane 90 degrees (vertical).
9. Click the Slice button under Slice plane. [*RESULT: The polygon is split in two*]
10. Click the Slice Plane button again to deselect it.
11. Select the new polygon on the right to select it.
12. In the Modify menu under the Edit Geometry rollout click Extrude.
13. Click and drag up on the polygon to extrude out the garage.
14. Click Editable Poly in the stack to select by object.
15. Select the Box.
16. Squash the Box using the Non-Uniform Scale tool until Z reads 10 units.
17. Move the Box up until the bottom of the new Box rests on the top of the first Box.
18. Scale the top box until it slightly overlaps the base of the house.
19. In the Create panel click Box.
20. In the Perspective view drag a square 5 units in width and 12 units in length, then release the mouse.
21. Drag up until the Height reads 50, then click to finalize the Box.
22. Move the box into alignment with the right of the house to use as the fireplace/chimney.
23. Our house is now modelled.

Modelling A Simple Tower

1. In the Create panel click Cylinder.
2. In the Top view drag a circle with a radius of 20 units, then release the mouse.
3. Drag up until the Height reads 120, then click to finalize the Box.
4. Set the Sides to 7 and the Height Segments to 1.
5. In the Create panel click Sphere.
6. In the Top view drag a circle with a radius of 30 units, then release the mouse.
7. Set the Segments to 6.
8. Using the Select and Squash tool squash the Y axis until it reads 60.
9. Move the Sphere to the top of the Cylinder.
10. In the Create panel click Box.
11. In the Top view drag a square 3 units in width and 5 units in length, then release the mouse.
12. Drag up until the Height reads 45, then click to finalize the Box.
13. Move the Box to the top of the Cylinder object. [*HINT: This Box will be an antenna*]
14. Pressing CTRL, drag the new Box to the right make a duplicate.
15. Set the copies' height to 30.
16. The control tower is complete.
17. Add textures to the tower using the techniques learned earlier in this article.

Placing New Objects In Flight Simulator (e.g. FS2002)

1. Start Flight Simulator, and then start a flight.
2. Once the flight is running, RMB click and select Top-down to get to the top-down view.
3. Press the "Y" key to enter Slew mode, then go to the location that you want to place the building.
4. In the upper left-hand corner of the Simulator, you'll find the latitude, longitude, and heading of your exact location: record these values for future reference.
5. Open gMax and from the File menu select Export.

Build Scenery Using GMAX

6. In the Select File to Export dialog box, make sure the Save as Type box is set to Flightsim BGL file (*.BGL).
7. In the Save In box, navigate to C:\Program Files\Microsoft Games\FS2002\Addon Scenery\Scenery.
8. Type a name for the file, and then click Save.
9. Select the Scenery tab in the Make BGL options and type the latitude, longitude, and heading values from FS2002 and then click Start.
10. Back in Flight Simulator, press ALT, click World, and then click Scenery Library.
11. In the Scenery Library dialog box, click OK.
12. Your house now appears in the scene.

For more information on gMax and FS resources visit Discreet's Web Site at: [**www.discreet.com**]

You now have the foundation for creating Flight Simulator 2002/04 aircraft and buildings. You can use this knowledge to create your own environment and experiences. Have Fun!

By Danny Riddell
For TecPilot

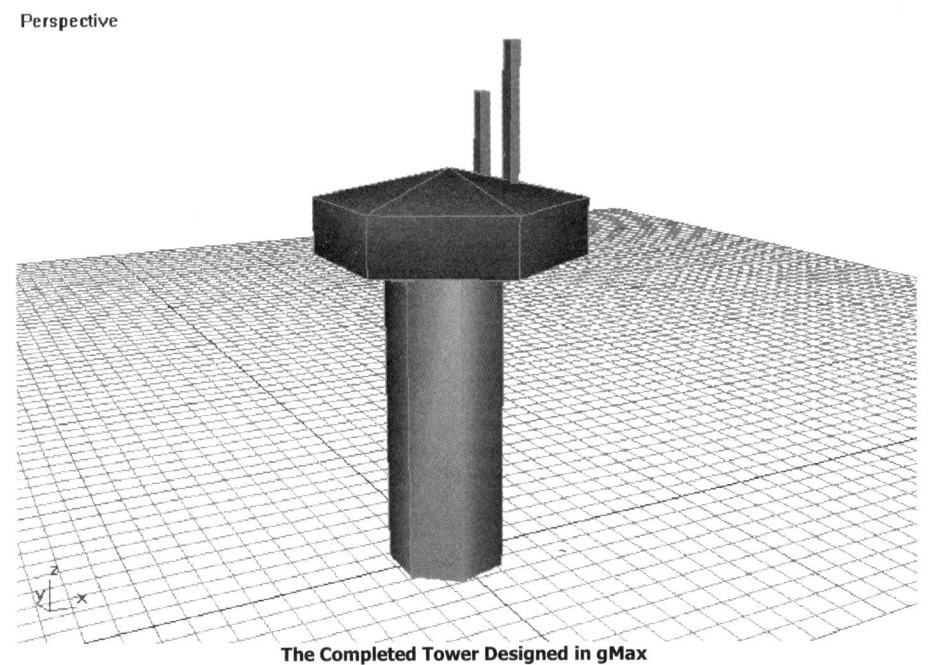

Perspective

The Completed Tower Designed in gMax

SECTION six

Navigating In Wind

Navigating In Wind

We are generally spoilt when we fly in the flight simulation world because we can control the weather and many people use this to give themselves perfect flying conditions. The wind in particular will be either turned off or set to a trivially small value, perhaps to make sure we get the particular runway we want from air traffic control. Let's be honest, real life is rarely like this. Wind is almost always present in the British Isles or elsewhere and even when it seems calm on the ground, there will probably be a significant wind at cruising level. The most obvious problem with high winds is the difficulty they present during takeoff and landing. Whilst winds can tax the pilot's skills during departure and arrival, the more important aspect is the effect that winds have on navigation. After all, there is no point in being able to perform a perfect landing in the worst conceivable cross wind if you cannot find the airport in the first place.

To be able to navigate successfully, you need to be able to start with your route and work out the headings you will have to steer. This can only be done properly if you make the correct adjustment for the wind. Before you can do this, you need to understand two different types of heading, because if you use the wrong one, you may end up miles away from your intended destination.

The North Pole is generally regarded as the point at the very top of the world through which the world's spin axis passes. Unfortunately the northern magnetic pole of the world is in a different location to the North Pole. The same applies at the southern end of the globe too. As a compass always points to the magnetic poles, it will not normally point to the actual North Pole. There will be some places in the World where it does but this is coincidental, only happening where the North Pole and the Magnetic North Pole are lined up. However for almost everywhere on the globe, there will be a difference. The compass in your aircraft is like any other compass, which means that it points towards the magnetic poles (or the magnet inside it does). This is why it is called a magnetic compass. It may look different to the compass used for walking in the countryside because it has a dial but it is essentially the same thing. Now a pilot will always align the Direction Indicator (DI) with the magnetic compass and hence the reading given on the DI is also relative to magnetic north.

Maps, on the other hand, are always plotted relative to the actual North Pole. The lines of longitude shown on a map run directly to the North Pole. This means that there is lots of potential for confusion between whether a bearing or heading is based on the North Pole or magnetic north. To avoid this potentially dangerous confusion, the two types of bearing are given different names. The direction of the North Pole from anywhere in the world is called True North. The direction directly to the magnetic north pole is referred to as Magnetic North. All navigational angles in aviation are measured against these and are referred to as either "magnetic" or "true" angles. In Great Britain, the difference between these is about 5 degrees and this is referred to as the "magnetic variation". If you look in some flight guides, or Jeppesen aerodrome plates you will see that each airport has its variation printed on the page. Variation is categorised as either "east" or "west" and in the British Isles they are "west", which means that the Magnetic Pole appears to be to the west of the North Pole. The amount of variation changes significantly throughout the British Isles and varies even more if you start to travel greater distances. You therefore need to check the variation at several places along your route. You cannot assume that it will be the same at the departure and destination airports.

The second difficulty with the simple task of finding the heading to fly is associated with the age-old discovery that the world is not flat. Life would be simpler if the World was flat, not just because flying over the edge and looking underneath would be amazing but also because it would be easier to produce maps. If you take a flat piece of paper and try to wrap it tightly around a spherical football without creasing or folding the paper, it is impossible. That's what a map is effectively trying to do, show a spherical object on a flat piece of paper. However you try to do it there will be distortion and the mapmaker's task is to minimise that distortion. For small areas the distortion is negligible, however even for an island as small as Great Britain, one degree of longitude is about 39 miles in the South and about 31 miles in the North. Despite the skill of the cartographer, the effect is clearly visible and if you look at an aviation map of the UK, you will see the lines of longitude and latitude converging and curving throughout the map. So how can you hope to plot an accurate course? Aviation maps that are designed for navigation are

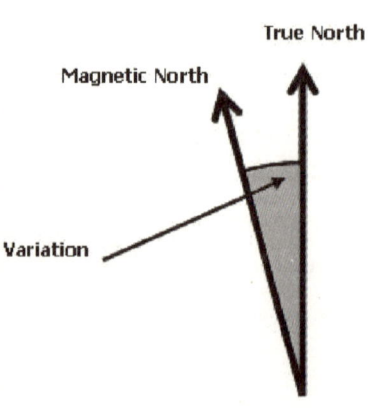

Navigating In Wind

typically produced using a "Lambert Conformal Conic Projection" or "Transverse Mercator Projection". Most pilots will recognise these names but will not remember which is which because it is not really necessary. The clever stuff has been done for you by the map developers. Maps produced using these projections have two very useful properties. Firstly, distances in any direction can be measured reasonably accurately from the map and secondly great circles, which are the shortest distance between two points on a sphere, are represented as almost straight lines. This is good news because the pilot can draw a line on a map and then measure the direction and distance accurately enough to use it for flying.

What you do need to know about, however, are rhumb lines. A rhumb line is a line that crosses each meridian of longitude at a constant angle. These are what are normally flown over short legs of a hundred miles or so. To fly great circles you would have to constantly change heading. Most pilots want to navigate a leg by setting a single heading and then flying it, or in other words flying along a rhumb line. As most legs flown by pilots are relatively short, the difference between a rhumb line and a great circle is so small that it is not worth bothering about.

Plotting Your Heading

How should you determine the heading to fly? The first thing to do is to find the True Track, which is the direction relative to True North from one waypoint to the next. You can get this in a number of ways. The most common way for pilots flying VFR (Visual Flight Rules - i.e. not instrument flying) is to draw a line on a map between the two waypoints. A protractor (ideally one designed for aviation) is then placed over the route half way between the ends and the angle clockwise from True North is measured. The lines of longitude provide an easy reference for lining up your protractor with True North. Although this is ideally done on an aviation topographical map, you can do this reasonably well on any map that has true north on it.

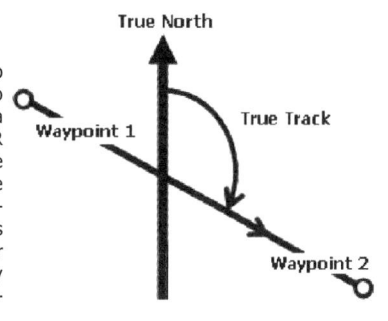

Now you have the true track for the leg, the next thing to do is to measure the distance between the two waypoints. You won't use this immediately but you may as well measure it while you have the map in front of you. Most maps have a scale on them to make this easy, although pilots use rulers that are calibrated specifically for their aviation maps.

A main planning objective is to have a heading that will fly you directly from waypoint 1 to waypoint 2. There are two more basic pieces of information you need. They are the true air speed (TAS) at which you intend to cruise, together with the wind speed and wind direction. Start by considering the TAS. This is based on the indicated airspeed (IAS, which is the speed you see in the cockpit on your airspeed indicator), your pressure altitude and the temperature of the air. The calculation is sophisticated but is simplified for pilots by use of a so called flight computer, which either looks like a pocket calculator or an odd looking circular slide rule. If you have one of these devices, follow the manufacturer's instructions. Otherwise you are best to approximate the true airspeed by adding 2% to the indicated airspeed for every thousand feet of pressure altitude. Pressure altitude is the height shown on your altimeter when the altimeter setting window is set to 1013 mBar (or 29.92 in Hg). Hence if you are flying at a pressure altitude of 5,000 feet at 120 knots IAS, you should add 10% (5 x 2%) making your TAS 132 knots.

The wind "velocity" comprises both a speed and a direction. The direction should be given as a heading true, which is how weather forecasts are normally supplied to pilots on the ground. You must never ever mix true and magnetic headings. Again if you have a flight computer, this device will make the calculation relatively easy for you. If you do not have one you will have to adopt one of the following methods. First we shall look at the graphical approach.

Navigating In Wind

Start by drawing a line on a piece of paper in the true direction you need to travel between the two waypoints. This is the True Track and it is normally signified by two arrowheads drawn on the line. Next, add the wind at the start of the Track, drawing it in the direction it would blow the plane. The length of this line must be scaled directly from the wind speed. Hence if the wind is 20 knots, you could make it 20 mm long (1 mm equals 1 knot). The following diagram shows a south westerly wind added. Note how a wind is named by where it is blowing from and not where it is blowing to. The wind is normally drawn with three arrowheads.

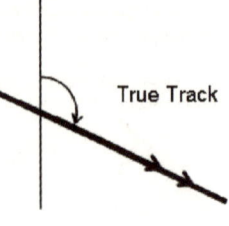

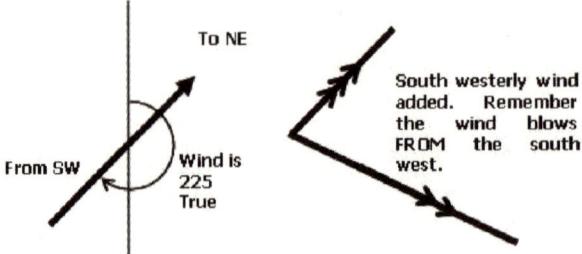

Now for the critical part. Draw a line for your TAS using the same scale that you used for the wind speed, which begins at the end of the wind vector and ends somewhere on the track line. So if your TAS is 120 knots and your scale is as before (1 mm equals 1 knot), the line would be 120 mm long. This could be done with a compass but it is not necessary and a ruler will suffice. This flight vector line is marked with a single arrow head.

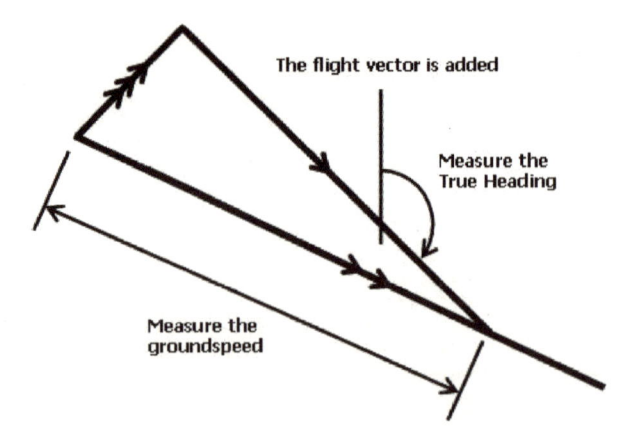

From this vector triangle you can now measure both the true heading that you will need to fly and your groundspeed, again using the same scale that you started with. It should now be obvious that the more accurately you can draw this vector triangle, the more accurate your final heading will be.

You may be thinking that you already knew that with the wind from the right, you would have to turn right a little to offset the wind. This is true and if you are prepared to determine logically which way you have to turn, you can quickly calculate both how much drift you have to apply and your groundspeed. For the mathematician, this is based on the "Sine Rule", which states that for a triangle:

A/Sin(a) = B/Sine(b) = C/Sine(c)

Navigating In Wind

This is for background interest only and you won't need to use this formula directly. If it means nothing to you, simply read on, the important parts come next.

Start by determining the angle between your track and the wind, we'll call this the wind angle here.

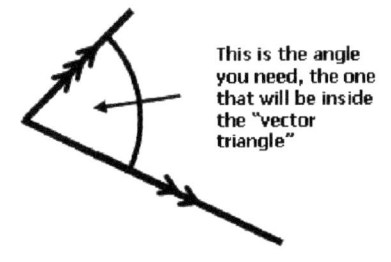

This is the angle you need, the one that will be inside the "vector triangle"

The formula is then:

DRIFT = ArcSine(Sin(WindAngle) x Windspeed / TAS)
and
GROUNDSPEED = TAS x Cos(Drift) + Windspeed x Cos(WindAngle)

The drift has to be either added or subtracted from your track true to give you your True Heading. In the example shown, the wind is from your right, so you would have to turn right to compensate. Therefore you would add the drift to your heading. If this calculation worries you, stick to the graphical method, however if it does not, you may find it quicker, not to mention more accurate.

When you fly an aircraft, you fly using a Magnetic Heading, all that remains is to convert the heading from True degrees to Magnetic degrees. If you look at any navigation map you will see the "magnetic variation", which we have already seen is the difference between True North and Magnetic North. You will also find it given in some but not all, flight guides. However, far and away the easiest place for TecPilot members to find the magnetic variation is on the TecPilot web site in the airport data section. Find an airport close to (or on) your route and round the variation there to the nearest degree. Variation is given as either magnetic degrees west or east. You will need to add any variation west to your true heading. Conversely if it is an east variation, you will have to subtract it from your heading.

So now you know the Magnetic Heading you will have to fly, the other two key pieces of information that you need are the length of time the leg will take and the amount of fuel you will need. The duration for the leg is easily found by dividing the distance between the two waypoints in nautical miles by the groundspeed in knots and you have already worked this out. Your answer will be in hours, so you will also need to multiply it by 60 to turn it into minutes. If you have measured the distance in statute (normal) miles, divide it by 1.152 to convert it to nautical miles before calculating the leg time. By way of example, if your leg is 20 nautical miles long and your groundspeed is 120 knots, the time taken will be:

60 x distance / groundspeed = 60 x 20 / 120 = 10 minutes

You will probably end up with figures more like a distance of 24 miles and a ground speed of 137 knots, giving a time of 10.51095 minutes. When you get answers like this, round them to the nearest minute. In this case you would use 11 minutes.

The fuel calculation is not much harder. The fuel required for the flight is straightforward and calculated by multiplying the fuel consumption by the time for the flight. As fuel consumption is usually given in gallons per hour, you will need to divide your total flight time by 60 too. Hence if your total en-route flight time is 134 minutes and the fuel consumption is 7 gallons per hour, your fuel used would be:

(Time in minutes) / 60 x (Fuel consumption in gallons per hour) = 134 / 60 x 7 = 15.6 Gallons

Navigating In Wind

Next, you need to add fuel to cover start-up taxi, climb, descent and shutdown to this. Ideally you would use data based on the published aircraft / simulator manual. As this is often not available, you may have to carry out some flight testing to determine figures for these. The alternative to this is to add some time that will consume roughly the same amount of fuel. Add ten minutes to allow for start-up taxi and shutdown, then add 15 minutes to cover the climb to cruise altitude and finally add 45 minutes emergency reserve. This gives an additional 70 minutes, changing the fuel calculation to:

= (134 + 70) / 60 x 7 = 23.8 **gallons**

Even this is not enough, you should also allow sufficient fuel to divert to your alternate (by planning the route and the fuel that will be consumed to take you there). Be clear that diversion fuel is not the same as your emergency fuel, you should be planning to land with that on board even if you have to divert. The emergency fuel covers things like holding while a runway is cleared after a plane has crashed on it, or unexpectedly high fuel consumption.

There is one more thing that you will need to add. It is possible that some of the fuel in the tank is not usable. This may be because of the shape of the fuel tank or the possibility of sucking air into the engine below a certain level, for example when the fuel is sloshed around in turbulence. If this were four gallons, you would need to have at least 28 gallons (rounded up) in your tanks for the journey. That's four hours worth of fuel for a two and a quarter hour flight. It may seem a lot but this is reality and the pilot has a legal responsibility to ensure that there is sufficient fuel on board for the flight. Despite this legal requirement, the fear of a hefty fine is not the normal reason why a pilot ensures there is sufficient fuel on board, it is the consequence of the crash after running out of fuel that make the pilot round up and then add a bit more.

If you are flying aircraft on public transport flights you would have to have your fuel planning policy approved, but it would typically include enough fuel for the following:

- **Taxi**
- **Take off and departure**
- **Cruise along the route itself**
- **Approach and landing**
- **Fly a missed approach**
- **Divert to an alternate airport**
- **Hold for 45 minutes**
- **Approach and landing at an alternate airport**

In addition to these there would be an additional reserve of fuel for the unplanned events as before.

Once you have all of your flight planned and all your calculations are complete, you will be in a position to go flying. If you are new to navigation you may be wondering how accurately you will need to fly. The answer is easy. You should aim to fly your heading, altitude and speed exactly as planned. Your heading should be within a degree or two at all times, which means that you should watch your direction indicator like a hawk. You will also need to keep your speed and your altitude accurately. This may seem overly exacting and in good visibility when you are proficient at map reading, you may find that you can get away with less accurate flying. Why so accurate then. For each degree that you are off the correct heading, you will drift one mile off course for every sixty miles flown (actually 57.3, but this is always rounded to 60). Hence if you flew a thirty mile leg but were flying off your correct heading by 10 degrees (perhaps because you had not checked your direction indicator against the magnetic compass), you would miss the waypoint by five miles. If the visibility were five miles, which is not unusual in the UK, you would sail past your waypoint without even seeing it. Fly at the wrong altitude and the wind will probably be different to the one used for planning. If things don't go according to plan, you need to know about it. That's one reason you have calculated the time that each leg takes. If the leg is calculated as being 10 minutes, you should be looking for your waypoint from about 8 minutes onward. If you don't reach it in 12 minutes, you can be confident you have missed it. If you don't fly at the correct speed, your timing will not work. This is not the worst part, fly at the wrong speed and the vector triangle will be wrong, meaning your heading will be wrong too.

Navigating In Wind

To give you some practice at flying in strong winds, here is a simple flight (**flight plan on page 87**) from airport to airport. The track and distance are given for each leg together with the headings. You should set the wind to 240/50 (50 knots from 240 True), which is a pretty strong wind. The visibility should be no more than 10 miles, which is actually quite typical for the British Isles. The recommended plane is the Cessna 172 or the Cessna 182, however, armed with the techniques above, you should be able to plan this flight for any aircraft and any weather conditions. It is left to you to decide where to collect any fuel that you may need to maintain your reserves. You may be tempted to begin with high visibility. If you do this, you will see that the headings and times work well. However, apart from being able to spot your waypoints from further away, the lower visibility makes no difference to navigating the flight. You will have to fly more accurately when the visibility is low but your waypoints should pop out of the haze at the right time as you fly between them. You could even add turbulence, which will make you concentrate harder but again everything will work out if you have planned it correctly.

The individual legs of the flight given get progressively longer and as a consequence they become progressively harder. Although it would be unusual to fly a single leg such as the longest one planned here without using some form of beacon, it is included to test your skill. Also, if you were flying cross-country in some parts of the world, such as Africa or Australia, there would be some massive expanses with little if any radio aids and few useful landmarks. The objective of this exercise is to develop your piloting skills, so most of the real work is left to you but to help you, here are some tips to guide you on your way.

After you take off from Southampton, climb on the runway centreline to 500 feet and then turn south, away from Popham and climbing to your cruising altitude. Then turn back to fly over the centre of the airfield. As you pass over it, turn onto your first heading and note the time. Record only the minutes part of the time, ignoring the "hours" part. The hour part of the time is usually obvious and is therefore omitted. Add the time for the first leg to this and put it into the ETA column. Maintain your altitude, heading and indicated airspeed accurately. When you have about 2 minutes left to run, start looking for Popham. It has two grass strip runways and is next to a dual carriageway. The layout and position is not quite the same as in the real world, however you should spot it easily if your navigation is good. As you pass overhead Popham, note the time in the ATA column and turn onto your next heading. Add the leg time to your ATA to give the ETA for Thruxton. Carry on flying in this way until you eventually reach Carlisle about four hours later.

Thruxton has a pair of crossed asphalt runways and you will need to be careful not to confuse it with Middle Wallop a few miles off to your left. Lynham is a large unmistakable aerodrome and you would have to be significantly off track to miss Filton. Turning for Gloucestershire airport, you will see the Severn and its bridge off to your left, with the next airport just past the end of the Bristol Channel. The flight on to Shobden is across some barren countryside, as is the track to Welshpool, so you will have to start flying increasingly accurately. Once at Welshpool, you will need to climb to the new cruising altitude to clear the peaks of Snowdonia dropping down again once you have past Caernarfon. As you approach Dublin, you should cross the coast just to the North of the peninsula that is to the north of Dublin Bay. The longest and final leg is now onward to Carlisle, which is by far the most challenging leg and should take you straight over the middle of the Isle of Man.

The final leg may give you some problems that are not necessarily of your making. Having both measured on a map and calculated the track mathematically (each several times), there is a high degree of confidence that the track and headings presented are as they would be in the real world. Despite this, you may find that the planned track will not take you directly to Carlisle on your simulator. This may be for a variety of reasons. There may be differences between the real world and the simulated world, which start to become significant over this distance. The magnetic variation used by the simulator may be different to that used for planning as it does change steadily with time. The wind may also be different to the planned value because of the way it is modelled. Nevertheless, with accurate flying, you should end up sufficiently close to be able to find Carlisle with a bit of visual navigation. The airport is on a radial of about 110 degrees from the end of the Solway Firth, although you will need to allow for the hefty wind. You could even call the airport as you approach it, ideally shortly before you reach your planned time if you cannot see the airport and you will hear where you are relative to them. Armed with this you can home in on Carlisle.

When you finally reach your destination, you will find getting the plane back on the ground in strong wind challenging. It is recommended that you land with no flap and at a higher than normal speed. However, as this is a flight simulator and you can try a short field landing with full flap. If you can control your speed sufficiently, you should be able to land and stop within the length of the aircraft!

Navigating In Wind

If you get lost along the route, don't just reset the flight, try to work out what you have done wrong. Check your actual heading and the time and they should give you a rough idea of where you are. Try to identify some ground features, or turn towards a nearby coast. If you are lost, you have to get yourself un-lost again. Once you know where you are again, try to fly directly to the next waypoint and continue your flight. Also remember that if you have set the wind incorrectly, all the planning will be wrong too. This is the equivalent of the forecast being wrong and it will have the most significant effect in strong winds, because your drift may be way out. A pilot would continuously monitor the plane's position and adjust the track accordingly, so if you have a map available to you, draw some lines on them and check your progress this way too.

You should recalculate the flight given here from scratch to practice the techniques. Don't worry unduly if your figures are the odd digit out, tiny differences are probably caused by rounding errors. What's more, despite checking, it is possible that there may be errors in the given plan. You, the pilot, are solely responsible for all aspects of your flight and therefore you must be confident in the data you use. The techniques given here will allow you to plan the flight with different winds and for different aircraft. You can even try different winds at different altitudes. Finally try creating some flights of your own, planning everything from start to finish. You now have the techniques to do this. Create and fly as many plans as possible and you will soon be in the perfect position to try this year's big challenge.

By Stephen Heyworth
For TecPilot

Navigating In Wind

Waypoint	Dist nmi	Alt feet	Track True	Wind / True	Vel kts	Hdg True	Drift	Var deg	Hdg Mag	TAS kts	IAS kts	GS kts	Time mins	ETA	ATA
Southampton															
Popham	15	2500	017	240 /	50	000	R17	4 W	004	116	110	147	6		
Thruxton	13	2500	274	240 /	50	260	R14	4 W	264	116	110	71	11		
Lyneham	23	2500	320	240 /	50	295	R25	4 W	299	116	110	96	15		
Filton	22	2500	272	240 /	50	259	R13	4 W	263	116	110	70	19		
Gloucestershire	27	2500	035	240 /	50	024	R11	4 W	028	116	110	159	10		
Shobden	34	2500	309	240 /	50	285	R24	4 W	289	116	110	87	23		
Welshpool	25	2500	337	240 /	50	312	R25	5 W	317	116	110	110	14		
Caernarfon	52	4500	304	240 /	50	282	R22	5 W	287	120	110	89	35		
Dublin	72	2500	287	240 /	50	269	R18	6 W	275	116	110	75	57		
Carlisle	152	5500	054	240 /	50	052	R02	6 W	058	122	110	172	53		

The Good Flight Simmer's Guide Mk.II - 2003

SECTION seven

Flying To Obtain Specs For Planning

Flying To Obtain Specs For Planning

When a new airliner is built these days, the design is so sophisticated and the simulations so good, that the test pilot's first flight is unlikely to bring any great surprises. It was not always so, the first flight of an aircraft used to be a genuine leap into the unknown. Although the expected flight characteristics would be known, the detail of how a plane actually performed would still have to be proven. Things like the spin handling would also be unknown until it was tried. This uncertainty was particularly true where the aircraft were pushing the boundaries of science at that time and even the test prototype of Concorde had an emergency exit just in case.

For the flight simulation pilot, flight testing holds some very strong attractions. It will develop your flying skills because you will have to control the aircraft precisely to gain sensible data. It will make you more aware of what is happening with your aeroplane, especially at the edges of the flight envelope. However, the biggest advantage is to be found when the aircraft you want to fly arrives on your hard disk with no instructions or other apparent indications about how to fly it. Test flying in these circumstances will give you all the data you need to fly your chosen aircraft with confidence.

Test Flying

Test flying aims to do a number of things. The most important decision to make is whether the pilots who will ultimately fly the plane will actually be able to fly it safely. This assessment should not assume that all pilots are superhuman. In any population there will be a range of skills from very good to relatively poor. By definition, half of the pilots will always be below average ability. The fact that the highly skilled test pilot can fly a plane is worth nothing. The objective is to ensure everyone who could act as pilot for the plane will manage it with appropriate ease. The aim is also to check that an aircraft is fit for purpose. An airliner should be very stable and docile, whereas an aerobatic aircraft like the Herr Walter Extra's would be useless if it handled like a Jumbo.

Another key factor is how well the plane 'talks' to the pilot. Does the plane let the pilot know how much further it can be pushed before it reaches the limit. An example of this is what happens when a plane approaches the stall, does the wing produce a noticeable buffet, or does the plane give no signal until it suddenly flicks, enters a spin and falls out of the sky. If the latter is true for a touring aircraft, a pilot concentrating on a map or flight guide could inadvertently enter a spin. At the other extreme there are planes like the Tomahawk that has a huge g-break at the stall, where both the nose and the wing can drop through about 60 degrees. However it does rattle and vibrate so much before the stall that even the doziest pilot would spot something was not quite right.

The test pilot is also part of a team sorting out any problems found. These will include everything from the ability to easily recover from a spin to whether the controls are too heavy (or too light).

Finally the test pilot is measuring everything in sight to confirm (or possibly generate) the data that all subsequent pilots will use to fly the plane. This is an area likely to be of great interest to flight simulation pilots. It will include determination of speeds for take off, cruise and landing, together with fuel consumption and so on.

How And What To Test

There are a whole host of things that would be tested in the real world that have no meaning in contemporary flight simulators, such as the stick forces at various phases of flight; position of the controls and the ease with which the pilot can reach them; or the engine reliability on runways submerged in deep rainwater. Test flying would also include proving the aircraft performance across the whole array of factors such as all possible flap settings, pressure altitudes, weights, wind speeds and so on. Once you have seen how to assess the basic parameters, you should be able to extend these to cover the full range of performance variables that will apply to your particular flying. We shall therefore stick to the basic elements that are useful to a flight simulation pilot.

When you have test flown a plane, unless you are the developer of course, you may find that the handling is so painful that you reject the plane completely and decide never to fly it again. On the other hand you may find a plane that flies so beautifully, that you forsake all others for months.

Flying To Obtain Specs Flight Planning

Handling Characteristics

The first things to consider are the general handling qualities of the aeroplane. Whilst these are mainly subjective judgements, they are nevertheless important aspects of the test flying of an aircraft. Some of them are immediately obvious to the pilot and a test pilot would be considering how much skill a pilot needs to fly the aircraft. If a plane requires exceptional skill to fly it, then it will be unacceptable for use by everyday pilots. On the other hand it may be highly appropriate for special purposes or specially trained pilots.

First you should consider the stability of the plane. Is it easy to control or is it a perpetual struggle? Is it easy to trim for level, climbing and descending flight? How does it stall? Is it imperceptible, docile or vicious? Assess all these with consideration for the intended use of the aircraft. An aerobatic aircraft would be expected to spin and flick easily, whereas a commercial jet should be as docile as possible. Whilst on the subject of stalling, is there plenty of warning before the plane stalls? The aircraft should somehow let the pilot know that it is about to stall, this can be anything from a stall warning horn and wing buffet, through to the stick shakers found on heavy jets. Even an extremely high nose attitude with the joystick miles back is a good indicator.

The next part of handling is about how effective the controls are. You should expect the plane to be suitably responsive to your control inputs. Passenger jets are typically far less responsive than general aviation aircraft. This is because of their size and because throwing passengers around is bad for business. One characteristic often found on older aircraft like Tiger Moth, Hurricane, Lightning and some kit planes is called "adverse yaw". If the plane suffers from adverse yaw, rolling left will also make the plane yaw right, which is the exact opposite of what you want it to do. The way around this is to use the rudder in conjunction with your roll inputs and keep the slip ball in the middle. Unfortunately some simulation planes have unrealistic (and probably unintentional) levels of adverse yaw. Another factor that is of interest is what you need to do with the ailerons in the turn. Some aircraft tend to roll back level if you let go of the joystick, others tend to roll even further the remainder will stay rock solid and not change their bank angle. None of these are 'right' but the first will give the most stable cruising plane.

Performance

The start of flight begins by definition with the take off. If the flight manual does not give a take off speed, you will need to work out your own. The rotate speed and initial climb speed are usually based on the stall speed of the aircraft. They are selected to give a reasonable safety margin away from the stall. The rotate speed (Vr) is normally ten percent higher than the stall speed in take off configuration and the initial climb speed should be twenty percent above the stall speed. Using this rule for a Tomahawk, which will stall at 49 knots, Vr would be 54 knots and the initial climb speed would be 59 knots. These compare very well with the book figures of 53 knots and 61 knots respectively. You can see that the rule gives speeds to a higher degree of accuracy than most people will fly and so many pilots would round these to more convenient figures, giving a more memorable 55 knots and 60 knots. These rules apply equally well to larger aeroplanes.

There is somewhat of a conundrum here, because you need the stall speed to find the take off speed but you must take off before you can find the stall speed. The difficulty is easily solved by finding a nice long runway, and looking for the minimum unstick speed. The minimum unstick speed will be sufficiently close to the stall speed to give an initial set of take off speeds. Find the minimum unstick speed by accelerating down the runway with the trim pulled way back to about the quarter (of its range) position. Hold the stick well back to raise the nose as early as possible but keep the planes attitude at about ten to twelve degrees. The plane will eventually fly itself off the ground and the speed at which this happens is the minimum unstick speed. You need to be careful here because some laminar flow wings may never lift off in this configuration and you will have to find the minimum unstick speed by repeatedly rotating at progressively higher speeds until you find the lowest speed at which the plane will lift off.

Next comes the climb performance. For a light aircraft, this is a case of timing the plane over a change in altitude and then calculating the rate and angle of climb. If you do this at several different speeds and then plot the results on a graph, it can be used to determine the best rate of climb and the best angle of climb for the plane. If you time a Cessna 172 climbing at full power between 1000 feet and 1500 feet you will get the first two columns of the table on the next page.

Flying To Obtain Specs For Planning

The Rate of Climb is then calculated by:

Height gained / time x 60

The angle of climb is calculated by:

ArcTan(ROC / Speed / 6082 / 60)

ArcTan is the inverse tangent, arc tangent or Tan-1 found on most calculators, spreadsheets or computer languages.

Plot these on a graph you will see that they clearly show the best rate of climb is achieved at 70 knots but anything between 65 and 75 knots would do. The best angle of climb, used for clearing tall obstructions, is at 50 to 60 knots. You might get a slightly different result if you tested the climb at every 5 knots instead of every 10 knots and ideally you would fly each test several times and then use the average result.

Strictly speaking, you should also repeat these tests for different weights (fuel loads) and at different altitudes and check for different results. However checking at different heights and weights is more important for bigger aircraft because of the huge ranges involved.

Speed (kts)	Time (seconds)	Rate of Climb (feet per min)	Angle of climb (deg)
50	43	698	7.84
60	36	833	7.80
70	35	857	6.89
80	36	833	5.87
90	41	732	4.59
100	61	492	2.78

**Cessna 172
Angle of Climb Performance**

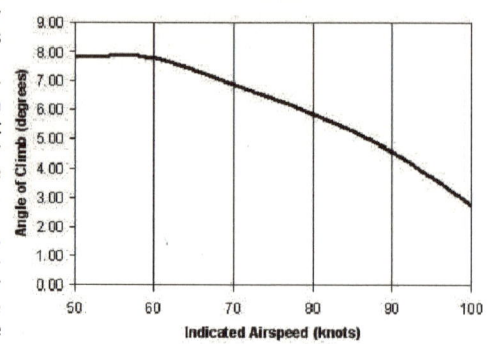

If you try the same tests with a larger aircraft you will have to time the climb over a larger height range. What's more you are likely to get some surprising results.

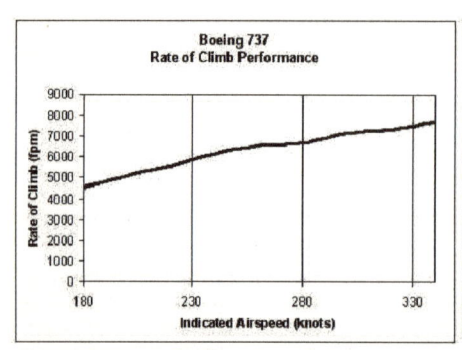

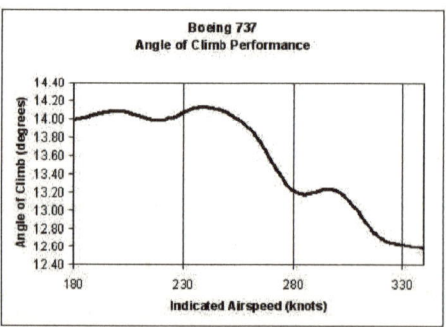

Boeing 737 Rate of Climb Performance

Boeing 737 Angle of Climb Performance

This shows that the faster you fly, the greater the rate of climb. You reach the speed limits before your reach the maximum rate of climb. It also shows a very wiggly curve for the angle of climb, however this is largely because of the small range (1.5 degrees) and the consequent sensitivities to small deviations in your measurement or flying. Nevertheless, useful data can still be drawn from the graph. The best angle of climb is at around 220 knots (you need to decide where the peak of the curve would be if you smoothed out the wriggles) and the best rate of climb is at the highest speeds. However, unless you are operating out of a very short field, the most important factor in climb for such jets is likely to be fuel economy.

Flying To Obtain Specs For Planning

Looking at the Cessna 172's glide performance, the data found by timing the descent over 1000 feet is as follows.

The rate of descent and angle of descent is calculated exactly as for the climb performance. When these are plotted on graphs, the lowest rate of descent can be determined as 650 feet per minute at 50 to 60 knots. On the other hand, if your engine has failed, you may need to cover as much distance as possible to reach some flat ground or a runway. If this case, the best angle of descent is at about 75 knots would be used.

Speed (knots)	Time (seconds)	Rate of Descent (feet per min)	Angle Of Descent (deg)
50	92	652	7.33
60	92	652	6.12
70	84	714	5.75
80	72	833	5.87
90	59	1017	6.36
100	48	1250	7.03
110	39	1538	7.86
120	33	1818	8.50

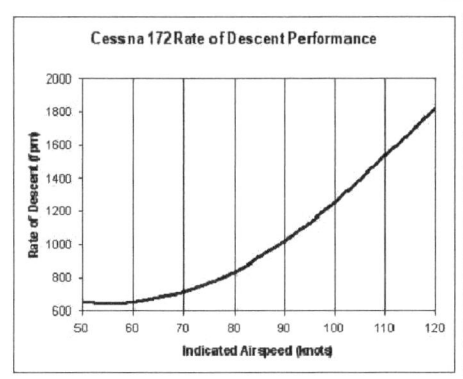

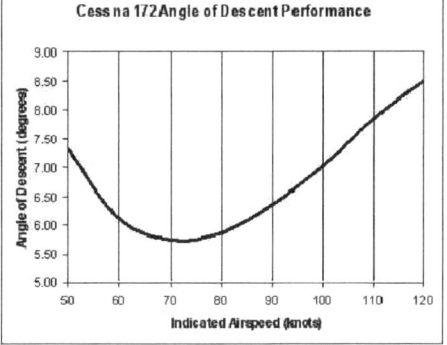

Finding V1

V1 is the decision speed for take off and is most applicable to turbine powered aircraft, including heavy jets. Once this speed has been reached, the pilot should continue the take off, even if there is an emergency. It the speed beyond which it is not possible to stop without running off the end of the runway and stopway. This speed is therefore dependent on the length of the runway. As we are flying a simulator, we can use Heathrow to do our testing. This is a superb place to do this test because the runway is long and because it runs close enough to East / West for the change in latitude to be ignored and the change in longitude to be used to calculate distances.

The first test we must do is to find how much runway is used when you need to stop after failure of one engine. Position yourself at the beginning of runway 09L using the menu to move you there. Set your take off flap, 5 degrees is recommended. Press Shift-Z (in FS2002) to bring up your location in longitude and latitude and make a note of them. Next apply full power and accelerate to 60 knots. At exactly 60 knots reduce one of the engines to idle to simulate your engine failure. The rules of the procedure say that it will take you two seconds to react. So two seconds after the engine failure apply full brakes and reduce the operating engine's throttle to idle. You can control each engine independently in Microsoft Flight Simulator by pressing "E" the pressing "1" to control "engine 1" and so on. When you eventually stop, make a note of the longitude. Next repeat the whole test from the same starting point but this time accelerate to 70 knots before you fail the engine. Keep repeating this until you have past your rotate speed. The difference between your longitude at the start of the test and the end of the test can be used to calculate the distance taken as follows:

(longitude change in minutes) x Cos(latitude in degrees) x 1854

The Good Flight Simmer's Guide Mk.II - 2003

Flying To Obtain Specs For Planning

As an example if you started your run at:
- *Longitude 0 degrees 29.32 seconds West*
- *Latitude 51 degrees 28.65 seconds North*

Simulated engine failure at 100 knots and stopped at:
- *Longitude 0 degrees 28.15 seconds West*

The calculation would be:

$$(29.32-28.15) \times Cos(51+28.65/60) \times 1854 = 1351 \text{ meters}$$

If you do this on a spreadsheet, which is advisable, remember that the latitude probably needs to be converted to radians before the Cosine is found using:

$$Cos(Radians(latitude \text{ in degrees}))$$

Now we need to move on and see what happens if we continue the take off after the engine failure. Use the same runway and settings you have just used for the accelerate-stop test but this time check that the altimeter is adjusted to the airport QFE and therefore showing your height above the runway of zero feet. Start by accelerating to 60 knots and then simulate the failure of one engine as before but this time continue the take off, rotating at the rotate speed Vr (which you have already worked out). Continue the climb to at least 50 feet above the runway. The real test flying is done using instrumentation and data recorders that your simulation plane will not have. All is not lost because there are alternate methods of data capture - the "instant replay" facility combined with running at quarter speed. These should enable you to note the longitude and latitude at the start of the test run, the longitude at take off and the longitude at 35 feet above the runway (that's why you set the altimeter to read zero on the runway). During this test, you may find some limitations with your simulated aeroplane. For example, you will probably find that, with one engine failed, the aircraft cannot take off above a certain flap setting.

As before you should repeat these tests with simulated failures every ten knots up to your rotate speed and calculate the distances as before. The distance from the start of your take off run to a height of 35 feet is the Take Off Distance Required (TODR). We shall use the distance to take off point later. If you now plot both the Accelerate-Stop distance and the TODR against the engine fail speed you should have a graph similar to that shown below, which was done for the Boeing 737 with full fuel and 5 degrees of flap. The X-axis has been renamed the Take Off Distance Available, or TODA, because this graph is used to check V1 against how much space there is to take off.

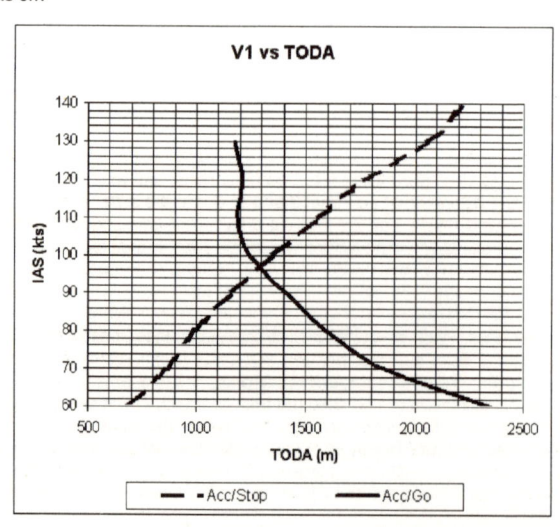

The Good Flight Simmer's Guide Mk.II - 2003

Flying To Obtain Specs For Planning

This is a very useful graph, as it allows you to select the V1 for any runway. If the runway is 2000 metres long, you can select a V1 anywhere between 67 knots and 128 knots. If you have an emergency, it's always better to deal with it on the ground than when you are in the air. So when there is an engine failure, it is normally sensible to stop if you can and you would be well advised to use 128 knots as your V1. The limit is reached where the line crosses at about 1300 metres. It is not safe to take off from runways shorter than this. For example, when taking off from a runway 1000 metres long, if the engine fails once you have accelerated past 80 knots, you do not have space to stop and you cannot take off. Even with a runway 1200 metres long, engine failure between 92 knots and 105 knots will leave you running off the end of the runway whether you try to brake or continue to take off.

Another way of looking at this is that if you have an engine failure at 90 knots you can only stop again on a runway longer than 1150 metres and could only take off on a runway longer than 1400 metres. Therefore if you were on a 1000 metre runway you could neither stop nor continue following the engine failure.

Landing Performance

For most pilots, there is a strong desire to have the same number of safe landings as they have take offs. A critical factor for safe landings is to be sure that the runway is long enough. To do this you will need to know the stall speed of the aircraft in the landing configuration. This is easy to test by flying the aircraft straight and level with all flaps and undercarriage lowered. Cut the power and hold your height as long as possible, when the plane can no longer hold its height, despite trimming backwards, this will be the stall speed. In many planes there is a distinct and sudden loss of lift combined with the nose dropping uncontrollably. In others it will be harder to detect the precise moment of the stall. The final approach speed, Vref, is 1.3 times this. So if the stall happens at 97 knots, Vref will be 126 knots. Some companies increase this a little and it would not be unreasonable to add 10 knots to the calculated speed for a heavy jet.

The Landing Distance required by an aircraft is based on the distance that a plane covers during a normal approach, landing and coming to a halt. The distance is measured from the point at which the plane is 50 feet above the runway until it stops. The stop is achieved using the brakes alone, without using any reverse thrust. The distance can be measured at Heathrow using the methods described for take off performance, which are calculated from the latitude and the change in longitude. The length of runway required is then calculated as 1.67 times the measured landing distance. For a Boeing 737, this was tested and calculated to be about 1350 metres.

Again you should repeat these take off and landing tests several times at a number of different weights (fuel loads), pressure altitudes, flap settings and with different head and tail wind components. However for the purposes of simulation, you will probably only need to cover a handful of configurations.

Efficient Performance

Climb performance is a compromise between a number of factors but at the end of the day it all boils down to a balance between the most economic performance, operating range and the time taken for the flight. One of the advantages for flight simulation is that you don't have to worry about the purchase and operating cost of your aircraft. On the other hand maybe you are worried about flight time and fuel consumption. Operate the wrong way and you will run out of fuel long before you reach your destination. What's more, for flight planning you are going to need to know what your fuel consumption is. Fortunately these are relatively easy things to test.

Starting with climb performance. Position your aircraft at Heathrow again, one of the 27 runways will be best as your climb will take several miles and this will take you away from Greenwich and the more complicated calculation if you pass from the Western to Eastern Hemisphere. Make a note of your position in Latitude and Longitude as before. This time, however, also note the fuel on board, which can be done either via the dials on the panel, or by checking through the menu system. The fuel on the menu system is a better, because it is both more accurate and shows the total fuel in all of your tanks. You will also need to use the aeroplane's built in clock if there is one, otherwise you will need a stopwatch. Decide on the climb profile you are going to test fly. Typical profiles you can test include, full throttle climbs at fixed speeds, fixed rate of climb speeds, maxi-

Flying To Obtain Specs For Planning

mum permitted speed climbs and in fact anything you wish. Each time you fly a profile you will need to create a table and fill it in as you go. To demonstrate this, the following table was used to collect data from a Boeing 737 climb at 500 feet per minute. The plane was throttled back to 250 knots below Flight Level 100 and then

FL	Time	Position (deg)	Position (mins)	Fuel gauge	Lbs Fuel	Flight Time (mins)	Dist Flown (miles)	Fuel Used (dial)	Fuel Used (lbs)
0	16:00	0	26.09	10	35593	0	0	0.0	0
50	23:10	1	11.97	9.6	34330	7.2	29	0.4	1263
100	32:58	2	24.06	9.2	33136	17.0	73	0.8	2457
150	42:43	3	49.04	8.8	31856	26.7	126	1.2	3737
200	52:35	5	21.50	8.3	30605	36.6	184	1.7	4988
250	62:36	7	1.18	7.9	29357	46.6	246	2.1	6236
300	72:16	8	48.55	7.5	28102	56.3	313	2.5	7491
350	82:10	10	42.57	7.1	26888	66.2	384	2.9	8705
400	92:03	12	35.21	6.7	25813	76.1	454	3.3	9780
430	98:00	13	42.52	6.5	25213	82.0	496	3.5	10380

increased to 280 knots. At about FL320, the plane was controlled to Mach 0.75 to prevent over speed.
The first six columns in the table above were collected during the test flight by pausing the simulation when the flight level was reached. The last four columns were calculated later. These are straightforward additions and subtractions with the exception of the distance, which is calculated using the following formula:

(longitude change in minutes) x Cos(latitude in degrees)

Which is very similar to the calculation done before.

Another test flight was flown. This time full throttle was used and the climb was increased to give 250 knots, 280 knots and Mach 0.75 as in the first test flight. This gave very high rates of climb at the lower altitudes. The results were then plotted on the following graphs to show the differences.

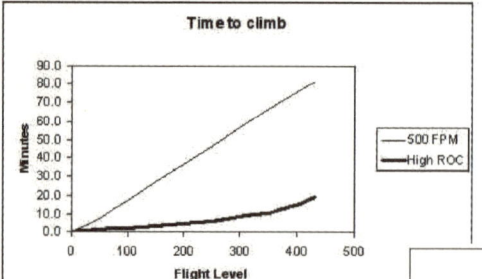

Not surprisingly the first graph show that when using the maximum climb possible it takes considerably less time climb. This took no consideration of the engines, which would probably object to the use of full power for nearly twenty minutes. As expected, it takes about one hour twenty minutes to reach forty thousand feet at 500 feet per minute.

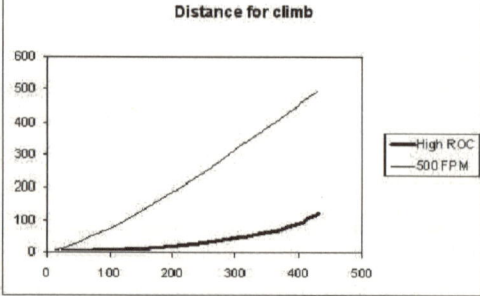

The graph showing the distance taken to reach the cruising level looks similar to the one for time but look at the distances involved. Although 500 feet per minute is the minimum required IFR climb rate, taking several hundred miles to reach cruising level in a passenger Boeing 737 would certainly raise a few eyebrows in air traffic control. Surely the maximum climb rate must be very heavy on fuel by comparison. Well, look at the next two graphs taken from the flight test and you may be surprised.

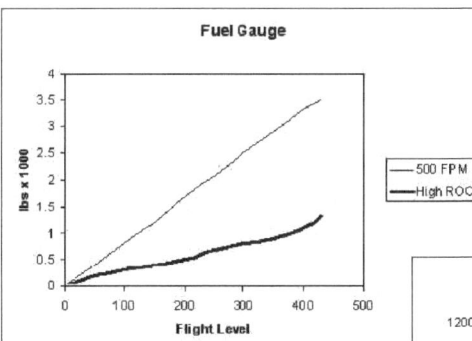

Flying To Obtain Specs For Planning

The fuel used to reach any cruising level is far less at the higher rate of climb. Admittedly that is only half of the story, because at FL430 you would still have another 375 miles to cover to reach the top of the slower climb. Even when you take this into account, the fuel used is about 40% less using the steeper climb.

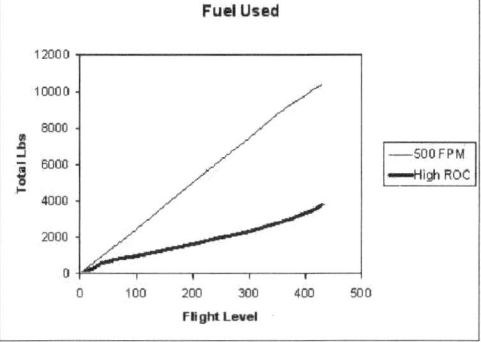

When you have flown several of such tests and compared them, most easily done using graphs such as those above, you will be able to decide on the most efficient way to operate your test aircraft. As a bonus you will also have some planning data for your aircraft. The graphs will enable you to read off the time and distance taken to reach any flight level you decide to use. You will also know how much fuel will be used in the process. Although this example was flown in the Boeing 737, the technique will work just as well on anything from a Cessna 140 to Concorde. All you need to do is make sure you choose suitable altitudes to take the data. For example, the Cessna would be better tested every thousand feet because there will not be many points on your graph if you use every 5000 feet.

Cruise Performance

Once you mastered testing the climb performance, the cruise performance will be easy to check. All you need to do is to fly at a number of flight levels and take data at the start and end of a timed run of about 10 miles. The easiest way to do this is to fly along a radial away from a VOR DME, starting about 20 miles away to reduce any errors caused by the fact that DME measures the slant distance. Make a note of the time, fuel contents and DME at the start and end of the run. You should then be able to create a table similar to the following:

FL	IAS (knots)	Time (seconds)	Fuel Used (lbs)	Distance Flown (miles)	Ground Speed (knots)	Fuel per Hour (1000 lbs)	Fuel Per Mile (lbs)
100	250	128	103	10.0	281	2.9	10.3
150	280	106	97	10.0	340	3.3	9.7
200	280	100	89	10.1	364	3.2	8.8
250	280	96	88	10.4	390	3.3	8.5
310	280	84	79	10.0	429	3.4	7.9
350	254	88	73	10.0	409	3.0	7.3
410	221	86	65	10.0	419	2.7	6.5

Flying To Obtain Specs For Planning

The reason the Indicated Air Speed changes above FL310 is because the limit then becomes the Mach number; Mach 0.75 was used for this test. Showing this graphically, you can see why jet aircraft are flown so high. The higher you fly, the further you can fly for the same fuel. Cruising at FL410 instead of FL150 will save you an enormous 35% of your fuel in the cruise.

You can use this graph (below), or the equivalent for your aircraft, for fuel planning. To cruise 200 miles at FL350 will take 7.3 x 200 pound of fuel, which is 1460 lbs. You can also read off your ground speed of 409 knots, so you would cover those 200 miles in about 30 minutes.

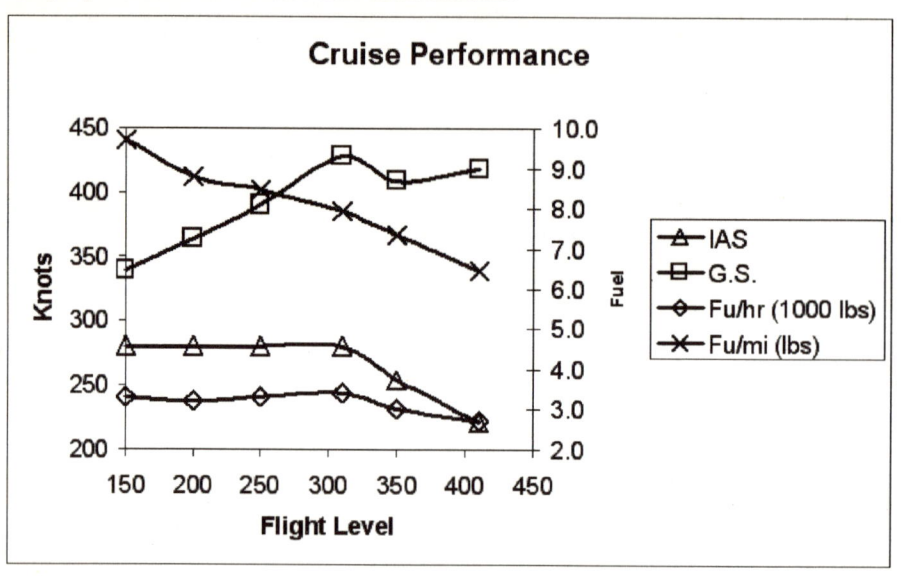

Descent Performance

Testing the performance of your aircraft in descent follows a similar format to the climb. Now for an airliner, you have already seen that the most efficient cruise is at the highest possible high level. This means that your most economic descent is a glide descent that starts as late as possible. You may wish to try some other descents, however you should include the glide descent in your programme of test. As you will stop this at 2000 feet, flying towards a VOR DME is ideal. For a glide approach in a 737 you should start about 90 miles away.

This time you will be logging your Flight Level, time, fuel, Rate of Descent and DME during your descent. From these you can calculate the distance to go and time to go. If you then plot the Rate Of Descent, Distance To Go and Time To Go on a graph you will get something like the one shown (right). The apparent glitch in the curves at FL100 is because of the need to slow to 250 knots. You should also plot the fuel used. However, as you are gliding all the way down, you will use very little fuel.

Putting It All Together

Once you have tested your plane, collected all of the data and then put it into both tabular and graphical form, you will have a sound understanding of your plane. You will not only know how it handles but also what the key speeds and limitations are. You will have all the information you need for your flight planning. Stringing together the climb data, cruise data and descent data, you should be able to calculate the time taken and fuel used for any flight at any altitude.

Flying To Obtain Specs For Planning

Glide Descent Data

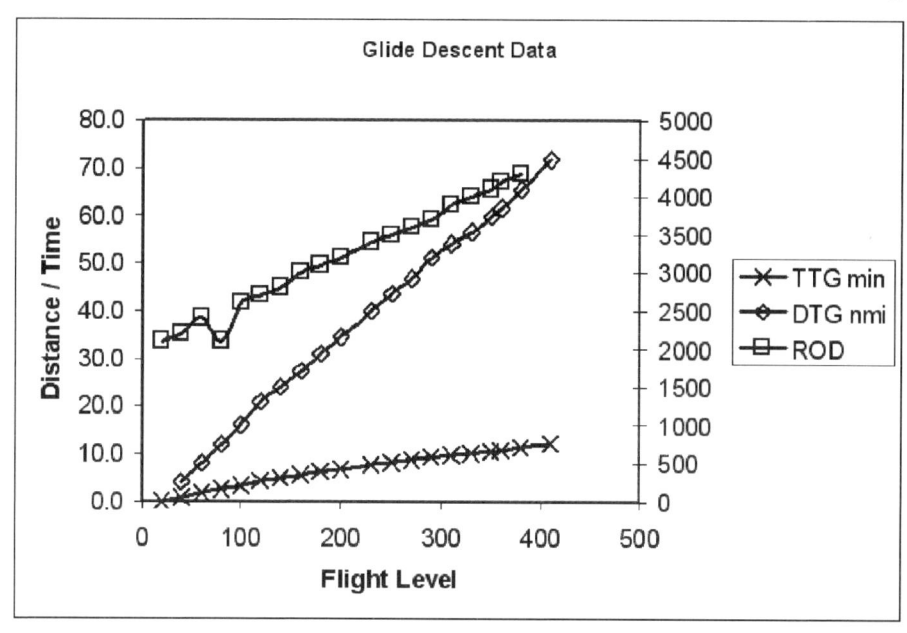

(Chart showing Distance/Time and ROD vs Flight Level, with curves for TTG min, DTG nmi, and ROD)

Plotting the graphs is not strictly essential if all of your flying is perfect but it does provide an easy way to check your data. Any large bumps in the curves may indicate either a misreading or less than perfect flying. When you do find bumps in the curve, you can either fly your test again or smooth them out by eye. The latter approach is useful, as one or two points that appear wrong on an otherwise smooth curve should probably sit on the curve anyway. However, you should be very careful, as there may actually be a good reason for the bumps in the graph. Plotting the data as graphs also enables you to take fewer measurements. For example, a graph that is compiled using measurements taken every 5000 feet can be used to read the data at intermediate altitudes. Some of the graphs shown here have many curves on a single graph and consequently the resolution suffers because the scale cannot be perfect for every line. Plotting individual full page graphs, each showing only one parameter, will result in graphs that are much easier to use for your flight planning.

The other thing to remember is that by repeating each flight test you will increase your ability to fly that particular test. Consequently the quality of your results should improve. The same improvement in the quality of your data will happen when you test several different configurations, such as flying at different aircraft weights. As a side benefit, you will probably be taking each plane to the extremes of its flight envelope that you have not visited before, which will both tell you more about how aircraft respond and broaden your flying skills. You will also discover any areas in which the plane misbehaves, which will be the flight regimes to avoid during your subsequent normal flight operations.

This has only scratched the surface of test flying and you can now see some of the reasons why test flying takes so long. For aircraft designers, both real and virtual, the test flights could also result in changes to the aeroplane's design. So not only is there the time taken to do the testing but there is also the time taken to redesign and then modify the aircraft. What is certain is that test flying any aircraft will give you far more information about how to fly it than you ever believed possible. So when you next open a box containing your new plane, or download something from the Internet, the first thing you should do is start test flying and collecting data. That way you will be able to carry out an informed assessment of the aircraft and know exactly how it should be flown.

By Stephen Heyworth
For TecPilot

SECTION eight

Flaps Explained

Flaps Explained

When aviation was in its infancy at the beginning of the last century, aircraft managed to take off and land without flaps. Initially, take off speeds and cruise speeds were similar. Then as speeds increased, techniques such as side slipping curved approaches were developed to land the tail draggers. Whilst such skills made for good pilots on simple aircraft, something was needed as the planes started to fly much faster. The answer was to fit flaps and slats to the aeroplanes, both of which perform a very similar function.

A Bit of Easy Theory

A little mild technical explanation of this graph is appropriate. Flaps and slats both affect the coefficient of lift (which can be thought of as much the same as lift in practical terms), however the way they do it is fundamentally different. The graph shows the relationship between the lift coefficient and the angle of attack. The angle of attack is the angle at which the air approaches the wing. When you raise the aircraft nose to maintain height at low speeds, what you are doing is increasing the angle of attack to get more lift coefficient and hence more lift. Flaps give even more lift by changing the shape of the wing; typically they turn a normal wing into one with higher lift and higher drag. Slats barely change the wing shape but they do keep the airflow attached to the aerofoil at higher angles of attack. They have little effect at cruise speeds, but as the plane slows down, slats allow a much higher nose angle before the wing eventually stalls.

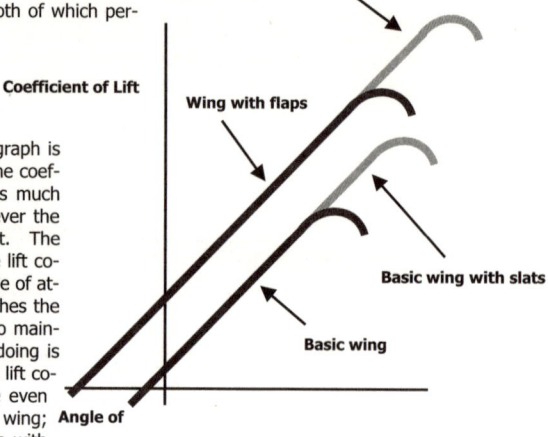

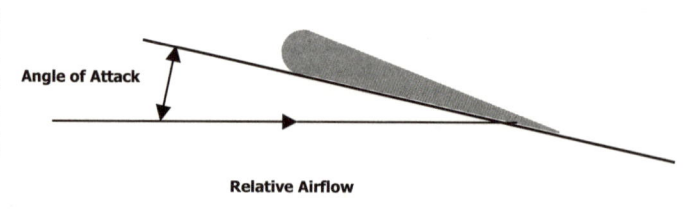

Increasing the lift generated by the wing allows a plane to fly more slowly. For many small planes the effect is relatively minor, changing the approach speed by about 5 knots. For airliners, however, the effect is both massive and vital. Cost conscious airline companies demand ultimate efficiency when flying in the cruise because they need to minimise the fuel used. This means that wings have to be designed to be extremely slippery at high speed. The unfortunate consequence of this is that they are useless at the low speeds necessary for landing. The solution to this conundrum is to fit flaps. Approach speeds are then much lower and the length of runway required is reduced. So instead of flying an approach at 200 knots it can be flown at 140 knots, much less challenging for the pilot and allowing shorter runways to be used.

The second effect of flaps is the addition of drag, which seems a strange thing to do for an aircraft designed to be as streamlined as possible. Unfortunately all that streamlining can be very embarrassing at times. Landing a slippery aircraft can be extremely tricky. Try landing the glider without using the airbrakes to see the effect; you lower the nose to loose height and the speed increases rapidly. Approach marginally too fast and you reach the other end of the runway before the wheels even get close to the ground. The additional drag from the flaps (or air brakes) rapidly bleeds off the speed during approach and landing. The other benefit of additional drag is that the engine is operating at higher power. This means that the gas turbines and turbochargers prevalent on modern aircraft will be spinning at a higher speed. If you then decide to go around, you will get a rapid response from the engine because the rotating shafts don't have to spool up, saving valuable seconds that you may not have.

Flaps Explained

The third effect of flaps is to change the profile of the wing, which in many aircraft results in a lower nose position. The beauty of this is that the pilot gets a very good view of the runway during the approach and landing. Trying to land without being able to see the runway is definitely a life shortening pastime, despite the impression you may get from several simulators.

The final major effect of the flaps is to deflect some of the slipstream from propeller aircraft downwards, thereby giving additional lift. This in turn gives some control of lift by changing the engine power, which is partly why the throttle should be used to control height during an approach to land.

Design of Flaps

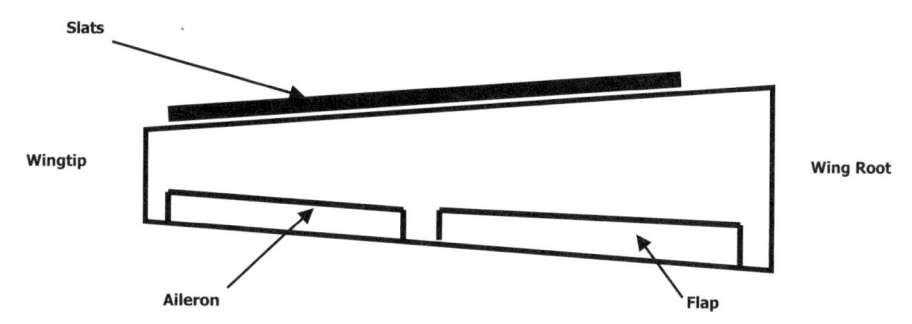

Flaps are usually positioned on the inner half of the back of the wing and are designed to droop downwards. There are several different types of flaps and even the most basic designs will add nearly fifty percent lift. In order of increasing lift, the types of flap you will find on most aircraft are:

(a) The Plain Flap - In which the rear of the wing lowers. This is the most common type of flap on light aircraft.	
(b) The Split Flap - Here the flap lowers away from the bottom of the wing but the top surface of the wing does not move.	
(c) The Fowler Flap - In this case the flap lowers and moves back. The gap between the wing and the lowered flap creates a slot, which dramatically improves the performance of the flap. This type of flap is found on some small aircraft. Fowler flaps can be seen in their most complex form on large aircraft, where many segments of flap extend from the back of the wing.	Simple form Complex form
(d) Kruger flaps - Flaps that extend forwards and downwards from the front of wings. They usually stretch across almost the whole span of the wing. They can be found in conjunction with fowler flaps on a number of modern jet aircraft.	

The Good Flight Simmer's Guide Mk.II - 2003

Flaps Explained

Slats are also fitted to the front of wings but work by ensuring that the air continues to flow across the upper surface of the wing at high angles of attack. The air is directed by the slot between the slat and the wing. The effect is disproportionately large when compared to their size. They are often found on older aircraft and STOL aircraft.

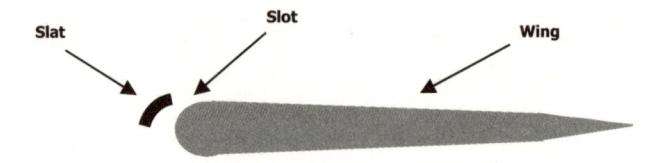

One final type of flap is worthy of mention, the **flaperon**. This is a combination between a flap and an aileron. In normal flight it is an aileron but when the flaps are selected, both ailerons droop but they continue to respond differentially to joystick input, allowing continued roll control.

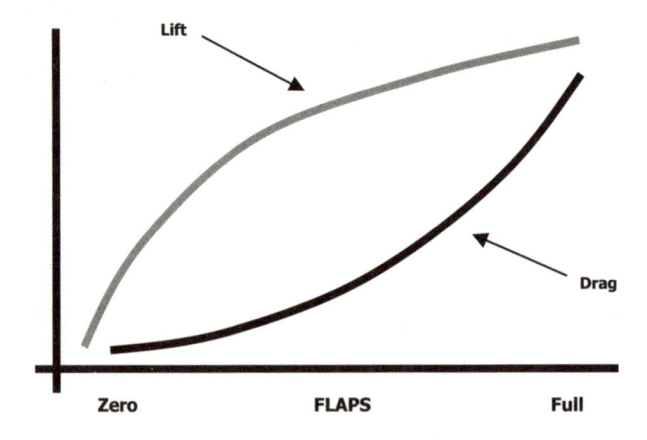

Most flaps can be lowered in stages and it is important to know that a small amount of flap will significantly increase the lift without adding much drag. Large flap settings have both high lift and high drag, so the later stages of flap add mostly drag. Hence lowering the flaps first produces mostly extra lift and then produces mostly extra drag.

When Flaps Are Used

There are three circumstances in which flaps are commonly used. Flaps are always designed for use during landings. If you consider the benefits mentioned above you will see why. They are particularly important for glide approaches and forced landings, where they are a way of guaranteeing a significant sink rate in the later stages of the approach. As a rule of thumb, you should always aim to use full flaps for landing.

Many aircraft use a small amount of flap for the take off too. This is very dependent on the design of the aircraft. The only way to tell the exact requirement is to read the aircraft manual. However as a guide most aircraft, other than small single engine propeller aircraft, use one stage of flap. Almost all aircraft will benefit from one stage of flap when operating from grass strips. Larger aircraft, which are designed primarily for speedy cruise, will need anything up to half of their flaps for take off. A large jet will typically use five or ten degrees for take off. The reason why you only use small amounts of flap for take off and climb is that you want the lift but not the drag.

Flaps Explained

The third circumstance that flaps are required is for reduced speed cruise. This could be when flying a holding pattern, or if the approach is slowed down by ATC. A low speed sight seeing tour or reconnaissance mission could be flown with some flap. In these cases only a stage or two of flap is normally used. In the same way as for the take off, the objective is to have increased lift without adding too much drag.

How To Use Flaps

There are two golden rules for using flaps:

(1) Only use the flaps within the permitted speed range. For many light aircraft the flaps cannot be used above a fixed speed. This is shown on the airspeed indicator by the white arc. If your speed is within the white arc, you are safe to deploy the flaps. As planes get more complex, each stage of flap will have its own limiting speed. The speed limitations are published in each aircraft's operating manual, however some panels have the limit speeds written by the flap leaver or indicator.

(2) Change the position of the flaps by one stage at a time. This applies to all planes, and the only exception would be during a forced landing, where you might even lower all the flaps in one go. The reason is that the plane will react to a change in flap position. The pitch, vertical speed, power required, trim setting and the control forces will all be different. It is essential to stabilise the aircraft between lowering or raising each stage of flap.

You should know that it is not good form to raise the flaps during an approach unless you are aborting your landing and "going around". If you need to raise your flaps during the approach, you shouldn't have put them down in the first place.

It is also sensible to start to lower the flaps well in advance of the landing. When flying a light aircraft in the circuit, you would normally put the first stage down either at the end of the downwind leg or at the start of the base leg. The final stage of flap should not normally be selected until you had received clearance to land. You should have your landing flap set before you reach 300 feet QFE.

If you need to abort a landing in a propeller aircraft, and you have full flap set, remove one stage of flap as soon as is reasonably practical. Otherwise you will have so much drag that climbing will be difficult. Try going around from the flare with full flap set and you will see how slowly the plane climbs.

If you are flying a glide approach or have an engine failure, always leave the lowering of stages of flaps until as late as possible. You should only lower each stage of flap when you are certain you can reach your landing point. It is far easier to lose height if you are too high than it is to gain height if you are too low.

The Downside of flaps

Every good thing has a downside and flaps are no exception. If you deploy your flaps at too high a speed, you will damage them. For some aircraft, the difference between approach speed and flap limiting speed can be very small indeed. This means that accurate speed control is essential.

Aeroplanes often have an inbuilt stability that ensures that its inner wing stalls before its outer wing. When the flaps are lowered the plane is less stable because the bulk of the lift is produced by the inner sections of the wings. The result is that the plane is far more likely to drop a wing in the stall. At the same time the ailerons are not working, so the roll can only be controlled with the rudder. The easy solution is to keep away from the stall speed, which you should do anyway.

Every time you change the position of the flaps, you have to reset the trim. Every time you change anything on a plane, you have to re-trim, so this should not be a surprise.

If you use the flaps for take off and the engine stops during the climb out, the extra drag will means that you lose speed at an alarming rate. You just have to be on the ball if this happens and lower the nose quickly and firmly.

Flaps Explained

The nose down attitude with the flaps lowered means that you will have to flair properly to prevent the nose wheel touching the ground first. This can be particularly crucial during engine failure at night when you cannot see the ground properly to judge the flair.

Cleaning Up

Flaps have played an enormous part in the development of flight. There has been a steady progression from the humble flapless aircraft in the early 1900's to the modern jetliners cruising intercontinental above vast oceans. These advances have only been possible because of the lowly flap. The current state of the art flaps may be found on military fly by wire aircraft, where the wings have active surfaces, that are essentially flaps, to make them more manoeuvrable. Or perhaps the state of the art flaps are those that allow a 400 tonne airliner to take off from a runway shorter than Hadrian's Wall. Flaps are now an essential part of flying so it is critical that you learn to use them properly. Their beauty is that they give the pilot so much and yet are extremely simple to operate when you know the few basic rules.

By Stephen Heyworth
For TecPilot

WWW.TECPILOT.COM

Join the online club for virtual aviators!

this is what you can look forward to...

- 30% Discount on Newly Released FS Products
- Indepth Reviews, and fascinating articles
- Meetings at Airshows, Exhibitions and Real Flight Simulators
- 2,000 Detailed Aircraft Specs
- 21,000 Detailed Airport Specs
- Real-Time Weather (Updated Hourly)
- Articles from people in real aviation! (pilots etc.)
- Regular News Updates
- BETA testing of products before they are released! (free software)
- A FREE Membership Card and Polo Shirt! (with Corporate package)

Come and join the team!   ...there's loads more besides!

Thousands of "Pilots" already have!

Look out for TECPILOT today... www.tecpilot.com

...We'll be happy to help!

SECTION nine

Gear Explained

Gear Explained

There are several milestones in a pilots flying career and one of them is the first solo flight in an aircraft with retractable undercarriage. As with most things on aircraft, the quality of the original design and the thorough maintenance regime will minimise the chance of the gear not descending when it's supposed to. Nevertheless, there always remains a "will it, won't it" feeling until you have seen the three green indicators light up. The other doubt in a pilot's mind is about forgetting to lower the undercarriage at all. After all, you can get away with occasionally forgetting to do many things on a plane and they are either instantly obvious or won't cause a disaster. Forget to lower your wheels before landing and you will have a very dangerous, not to mention expensive, experience.

Undercarriage can be generally split in two different ways. It is either fixed or retractable and it is either tricycle or tailwheel. There are other systems, such as the Harrier, which has two main wheels mounted on the fuselage one in front of the other, together with two wing tip wheels but these are the exception.

Tail-Draggers

The very first undercarriage consisted of simple skids but the limitations of this very quickly led to development of the tailwheel system. In these, there are two main forward mounted wheels and a single tail wheel or tailskid. Where a tailskid is used, the plane is often referred to as a tail-dragger. A simple consideration of the physics of the aircraft will show that the centre of gravity of the aircraft has to be behind the front two wheels and in front of the tail wheel, otherwise the plane would fall over when parked. In fact the centre of gravity has to be a reasonable distance behind the front main wheels to prevent the aircraft nosing over as soon as the throttle is opened. This positioning causes two of the three main problems with tailwheel aircraft. If the plane is landed without the tailwheel touching the ground at the same time as the main wheels, the inertia of the plane (acting at its centre of gravity) will push the tailwheel to the ground shortly afterwards. Unfortunately this accelerated and premature lowering of the tailwheel increases the angle of attack on the wing, which in turn gives the wing more lift and the plane takes off again. Unless the plane is landed in the perfect three point attitude, it will "bounce". It's not really a bounce in the true sense of the word but an immediate return to flight. It is possible to do a wheeled landing by extremely gently lowering the aircraft onto its main wheels at a much higher speed, and then letting the tail slowly drop but this is not the normal landing mode as it requires much more runway and precision.

The second problem caused by the position of the centre of gravity is ground looping. This is when the aircraft suddenly spins around and points in the opposite direction, often damaging a wing or the undercarriage in the process.

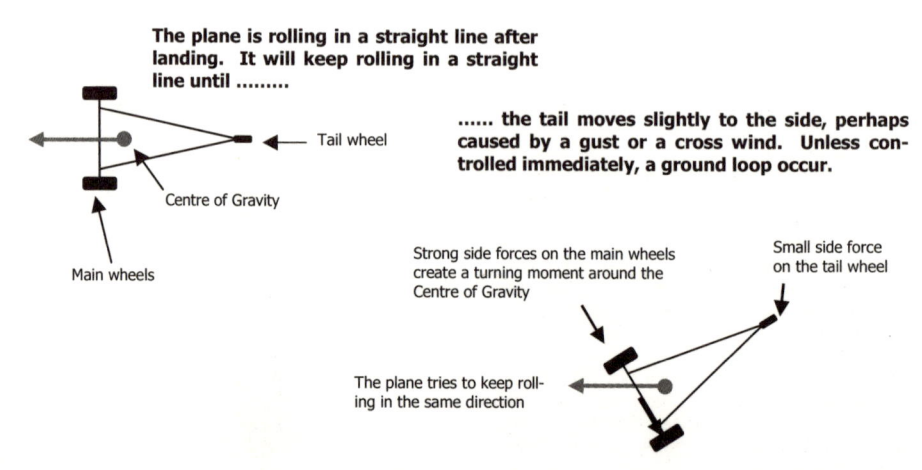

Figure 1 - What makes ground loops happen

Gear Explained

The laws of physics dictate that the centre of gravity of the aeroplane will always try to continue moving in a straight line. If the aircraft deviates slightly from rolling in a perfectly straight line after it has landed, as it inevitably will, the side forces on the main wheels, combined with the centre of gravity's desire to travel in a straight line, will result in the plane turning yet further away from the straight line. The more the plane turns, the greater the forces making it turn and once it has turned anything more than a small angle, the pilot has no chance of recovery. This is the ground loop and the pilot must be quick as greased lighting on the rudder pedals to prevent it. Many tail wheel aircraft have fully castoring tailwheels. This means that they are very manoeuvrable on the ground but the tailwheel is no help at all in preventing ground loops. Aeroplanes with skids or steerable tailwheels fair much better but the remedy is not complete. Ground looping is less likely to be a problem on take off because the engine is pulling at the front of the aircraft and blowing its slipstream over the rudder which gives far more rudder authority than on landing.

The other big problem with tailwheel aircraft is the forward visibility when on the ground. The low tail inevitably means a high nose, which blocks all forward visibility on most but not all, taildraggers. It is worst on older aircraft with vertically mounted engines or radial engines and has least effect on modern aircraft with horizontally opposed flat engines. The only cure for this is weaving from side to side and looking first to the left and then to the right of the engine cowling. This is why you will regularly see old war-birds like the Mustang swinging from side to side as they taxi.

Tricycle Undercarriage

The movement of the third wheel from the tail to the nose of the aircraft instantly removed all of these problems, with very few penalties. As a consequence the tricycle undercarriage arrangement has swept the world of aviation and tailwheels are now only fitted to some specialist types of aircraft, such as those designed for short field operation where the lower weight and drag combined with the three point attitude can be used to an advantage. The tricycle is much easier to handle on the ground because all the inertial forces will make it run straight. What's more there is also less drag during the take off run because the wings and tail plane are pretty much in line with the airflow. This can make the take off run shorter, allowing brisker acceleration. As a final advantage, because the propeller is rotating in the vertical plane, there are far less gyroscopic or "P" factor problems during the take off run, so the likelihood of veering off the side of the runway is dramatically reduced. All is not perfect for the nose wheel equipped aircraft because placing it at the front makes it very susceptible to high loads. Mishandling of the aircraft, especially during landing, can result in the nose wheel collapsing. Make them stronger and the plane becomes much heavier and there will be more drag. This is particularly a problem for low powered aircraft, where the extra drag and weight can have a significant effect on performance.

Retractable Undercarriage

Most small aircraft have fixed undercarriage rather than retractable landing gear. This is a compromise between cost, weight and performance. Whilst the clean lines of an aeroplane with retractable undercarriage mean that it has significantly less drag, this has to be offset by the increase in weight and cost. Where a plane is only capable of flying at low speeds, the weight and cost are normally dominant because the drag can be minimised. However, when the speed increases, the drag will increase roughly fourfold for each doubling of speed. This means that the drag of fixed undercarriage on any fast aircraft would be unacceptable. There are nearly as many landing gear systems as there are aircraft designs but those that retract generally have some common features.

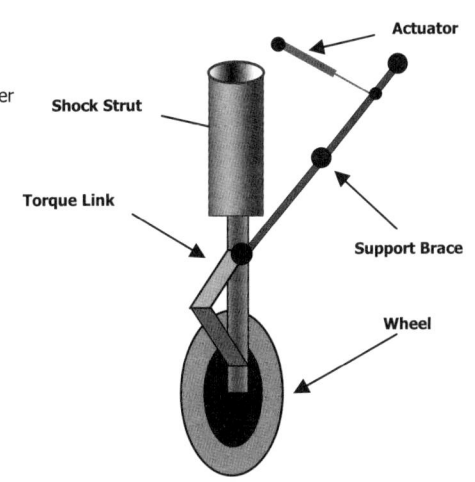

Figure 2 - The basics of a simple landing leg

The Good Flight Simmer's Guide Mk.II - 2003

Gear Explained

The landing gear leg usually swings downward to form a main vertical strut with a diagonal support brace. The main strut includes a shock strut that absorbs the landing impact and the bumps in the runway. The mains strut can be considered similar to an open (extended) bicycle pump; if you put one finger over the end and try to close the pump, it acts like a spring, as you try to close it, the pressure inside pushes back against you. Although a lot more complex than this, the shock strut works on this basic principle.

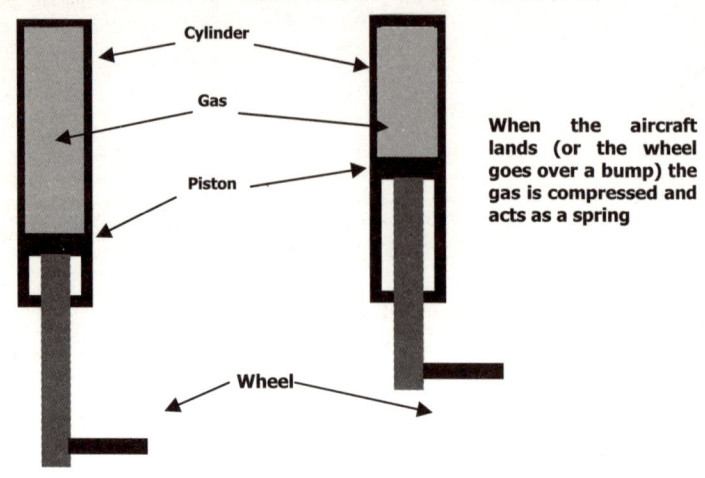

Figure 3 - principle of a shock strut

The diagonal support brace is to hold the undercarriage in the down position for landing. It does this by locking in position with an "over centre" mechanism. The most common example of an over centring device is a human finger. If you straighten your finger as far as you can (so that it curls slightly backwards a little) and press on the end, your finger will remain pretty straight. If you now bend your finger only slightly forward and press the end again, it folds quite easily into the palm of your hand. The principle is that the diagonal bracing strut folds out until it reaches a position where any force on the end cannot make it re-fold. A micro-switch is activated once the brace is in this position and it is this micro-switch that turns on the corresponding green indicator in the cockpit. Not every aircraft has an over centring device, however they should all have some form of locking mechanism that triggers the indicator in the cockpit when the gear is locked.

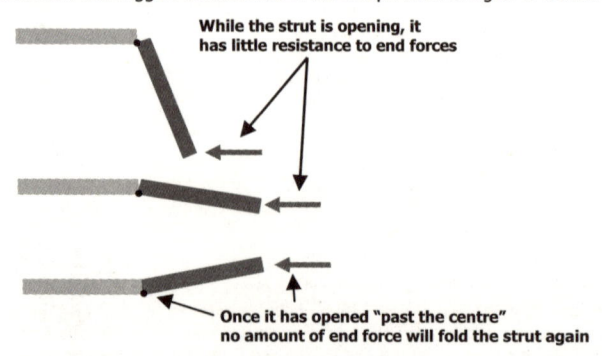

Figure 4 - A simple over centre device

Gear Explained

The third common elements in landing gear are the torque links. As the shock strut is in two parts, they can freely swivel relative to each other. This is no good, because the wheels need to be constrained to roll in the correct direction only. Two small hinged arms connect the top half of the shock strut to the bottom half, or connect the steering linkage to the bottom half of the strut.

Hinges

The torque link allows movement in the vertical plane, but does not allow any twisting movement. It therefore prevents the top and bottom of the shock strut rotating relative to each other.

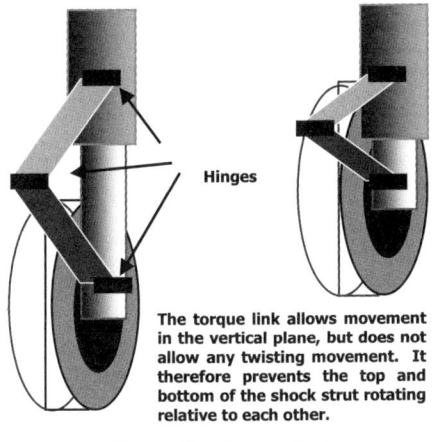

Figure 5 - Torque links

The mechanism for raising and lowering the undercarriage vary between the extremely simple devices with few components, to the multi-link systems that can twist and collapse the landing gear in many ways to make it fit into the limited space and shape available. Larger aircraft raise and lower the gear using hydraulic system, however smaller aircraft may power them with electricity, pneumatics or even the pilot's muscles. There is also usually some backup system for lowering the gear in an emergency and these range from secondary powered systems to simply letting the gear fall under gravity when the retaining catches are released.

Flying with Retractable Landing Gear

With the installation of retractable undercarriage comes the need for the pilot to manage it. We're all used to hitting "G" on the keyboard to raise or lower the wheels, however with increased realism, there is a little more that you should know about the subject.

Most aircraft will have a system to prevent you raising the undercarriage when you are on the ground. Again these vary from plane to plane. Some rely on the pilot taking two actions, for example raising the gear lever and moving an inhibitor lever. Most rely on interlocks, usually electrical "squat switches", that prevent the wheels retracting if they have weight on them. These squat switches can also activate a host of other systems including automatic spoiler deployment on landing and allowing thrust reversal to operate.

The pilot should ensure as part of the pre-takeoff checks, that the three green indicators show the landing legs are down and locked. Your walk-around would have already proved this, however you are really checking that the indicator system is working. It's better to find the fault on the ground than worry about it in the air.

The undercarriage can then be forgotten until after take off, when it is retracted. The objective is to raise the landing gear and get rid of its associated drag as soon as is practicable, because doing so will improve your acceleration and rate of climb. The time to retract the undercarriage in a single engine aircraft is classically when you have reached a point where you can no longer land on the runway in the event of the engine failing. It can be sensibly done at 200 feet with your other after take off checks and flap retraction. For larger aircraft, the undercarriage is normally retracted when there is a positive rate of climb identified by both pilots, however don't be too hasty with this because any measurements taken from sensors at the front of the aircraft will show a climb during rotation, which may be before the plane is fully airborne. Hence, waiting for at least 100 feet is wise.

As soon as you select gear up, the three green lights should go out and the red gear unsafe lights should come on. There may be only one red light to show that the undercarriage is neither lowered nor retracted, however the more sophisticated planes will usually have one per wheel. When the wheels are fully stowed away, the red lights will go out. What happens if the gear does not retract properly? It can be recycled, although it is often best to only do this once. If this fails to cure the problem it is best to land and sort it out on the ground.

The undercarriage is next called into play for landing and it is normally lowered as part of the pre-landing checks, which are typically done downwind or when approaching the airfield. For larger aircraft, the landing will normally be from an instrument approach. For an ILS, the glide slope is intercepted from below, and the gear should be lowered as soon as the glide slope needle starts to move. For other instrument approaches, the time to lower the wheels is on completion of the final turn.

Gear Explained

For straight in or vectored non-ILS approaches, lower the undercarriage when there are about 8 miles to go to landing. Checking that the gear is down again on short finals is always a sensible thing to do. It is unforgivable in flying to land with the gear accidentally up.

The undercarriage on every aircraft will have two critical speeds associated with it. There will be a maximum speed at which the undercarriage can be operated safely (Vlo), and a maximum speed at which the aircraft may be flown with the landing gear down (Vle). On some aircraft, the landing gear raising and lowering limit speeds are different, with the retraction speed usually less than the lowering speed. As these speeds all depend on the design of the aircraft, they can only be found in the aircraft manuals unless there is a placard near the lever. The undercarriage is usually raised at a low speed, even during a missed approach, hence this is unlikely to be a constraint as long as you don't forget to retract it when you should. However the undercarriage can be lowered and used as an air brake for emergency descents, or even just to slow a slippery aircraft down to approach speeds. Therefore it is useful to know the maximum speed at which it can be lowered. Once the gear is lowered, it is possible to increase the aircraft speed to Vle, which may be advantageous for an emergency descent.

The way that the gear is checked down on most aircraft is by the indicator lights moving through the red unsafe position to the "three greens" position. What happens if you don't get three greens? Well this will depend very much on the particular aircraft, but the first thing to do is normally to check that the indicator light bulb is working. On many aircraft this is simply a case of pulling out the offending indicator lens and lamp, and swapping it with another one of the three. If there really is a landing gear problem, it can often be cured by recycling the gear, so that the wheels are positively driven into the extended position again. The recycling may dislodge any unexpected debris or ice that had prevented proper deployment. The next option is to use the alternate system to lower the gear. This commonly involves winding them down manually. If all these fail, the next step is to talk to the controllers, burn off as much fuel as possible and then land with whatever you have got.

There will inevitably be some times when you are not happy with your approach and need to carry out the missed approach procedure, or the controller instructs you to go around. In these circumstances it is imperative that you remember to retract your undercarriage as soon as you have established a positive climb. As with the take off, your objective is to minimise drag and maximise rate of climb away from any solid objects.

As a final hint, remember that lowering or raising the landing gear on most aircraft alters the aerodynamic balance of the aircraft and consequently will require you to re-trim. Some planes manage to reduce this to the absolute minimum.

Whatever plane you choose fly, the undercarriage has the vital part to play. Some systems are a mechanic's nightmare, whilst others are extremely simple. The thing that really counts is that when your wheels do finally touch the ground, the undercarriage does its job and perhaps even accommodates the odd landing that is less than perfect. Although the fun part of flying is all about what happens in the air, the safe return to earth is usually seen as the most critical event by pilots and passengers alike. You will only be able to do this like a true professional if you manage your landing gear correctly. Happy landings!

By Stephen Heyworth
For TecPilot

WWW.TECPILOT.COM

Join the online club for virtual aviators!

this is what you can look forward to...

- 30% Discount on Newly Released FS Products
- Indepth Reviews, and fascinating articles
- Meetings at Airshows, Exhibitions and Real Flight Simulators
- 2,000 Detailed Aircraft Specs
- 21,000 Detailed Airport Specs
- Real-Time Weather (Updated Hourly)
- Articles from people in real aviation! (pilots etc.)
- Regular News Updates
- BETA testing of products before they are released! (free software)
- A FREE Membership Card and Polo Shirt! (with corporate package)

Come and join the team! ...there's loads more besides!

Thousands of "Pilots" already have!

Look out for TECPILOT today... www.tecpilot.com

...We'll be happy to help!

SECTION ten

Trimming Explained

Trimming Explained

A pilot should always fly the aeroplane by moving the joystick and rudder controls into the correct position. The forces that have to be used to do this can be significant, tiring and distracting. The fundamental purpose of trimming is to remove these control forces so that the pilot does not have to continually apply pressure. If trimming is done correctly, the pilot can let go of the controls for lengthy periods without the aircraft deviating from its intended course. What's more, when the plane does begin to drift away from heading or pitch, only very light forces are required to bring it back. As you can see, trimming is all about making accurate flying both easy and pleasurable.

The trimmer most frequently found on aircraft is the pitch trim, so we shall consider this first. To achieve continuous level flight, the pilot must hold the joystick or yoke in a fixed position. Unless the pilot is very lucky, there will be a force trying to move the joystick away from this position. If the stick is released, it will move and the plane will then either climb or dive. A pilot trying to resist this force for any length of time would quickly tire. If the forces were great and sustained, the pilot would ultimately lose control of the aircraft, which is generally not recommended. To prevent this, the pilot "trims out" the forces. The trim wheel is wound one way or another until there is no longer a force on the joystick. It is essential to realise that the joystick has not changed position. All that has happened is that the forces have been removed. The pilot can now let go of the joystick and the plane will continue in its level flight.

As planes get bigger and more sophisticated, they gain the ability to be trimmed in roll and yaw too. In essence the procedure is the same, hold the joystick for roll, or rudder for yaw, in the required position and trim out the forces with the appropriate wheel until there is no movement when hands and feet are taken off the controls. As planes get more sophisticated, power assistance is added to wind the sensitive trim wheels slowly through their large ranges. The correct order for trimming is to trim in pitch, then roll and finally yaw (rudder). Once a plane is fully trimmed, the pilot can get on with other essential tasks like map reading, tuning radios, etc., without having to concentrate all the time on controlling the aircraft.

Trim wheels are usually located in the centre console or underneath the centre section of the instrument panel. The wheels always move in the same sense as the help you want to receive. So if you are continually pulling back, you will need to roll the trim wheel back until the stick force disappears. If you have to push the stick right all the time you would have to move the roll trim to the right.

Trimming Simulators

How does this relate to flight simulators? Well, a slightly different technique has to be used when trimming virtual planes. This is because the joystick's springs are fixed, which means that the position the joystick moves to when you let go (its centre) is always the same. It is not changed by the trimming, which takes place in the software. In theory it is possible for a force feedback joystick to change its sprung centre but for most joysticks this is impossible. Your technique therefore has to accommodate this.

You should always begin by flying the plane without trying to trim it. Only when everything is steady, should you begin to trim. Starting with the pitch trim, you should feel whether you are pulling back or pushing forwards. If you are pushing forwards give a short burst of forwards trim and vice versa. It is absolutely essential that you continue to control the plane with the joystick at all times; you should never release the joystick and try to control the plane with the trim. If you still need to push the stick in the same direction, you should give another short burst of trim. Each time you do this you will have to move the stick very slightly to maintain control of the aircraft. As the forces get less, you should make your bursts of trim adjustment shorter. Eventually you will reach a position where you overshoot and have to apply a joystick force in the opposite direction. At this stage you should make the smallest possible changes in trim until you can let go of the stick without the plane pitching up or down. Beware though; it is impossible to trim many older simulators perfectly. Having trimmed in pitch, you would then trim in roll and rudder in exactly the same way. Nine times out of ten you will only need to trim in pitch unless you drain all the fuel out of one wing but you would never forget to change tanks, would you?

As you get more experienced you will gain a feel for when to apply large handfuls of trim and when to be exceedingly delicate. You will also learn that some actions always require a particular change in trim. For example, lowering the final stage of flap on approach will always require about the same trim change.

Trimming Explained

How Does Trimming Work?

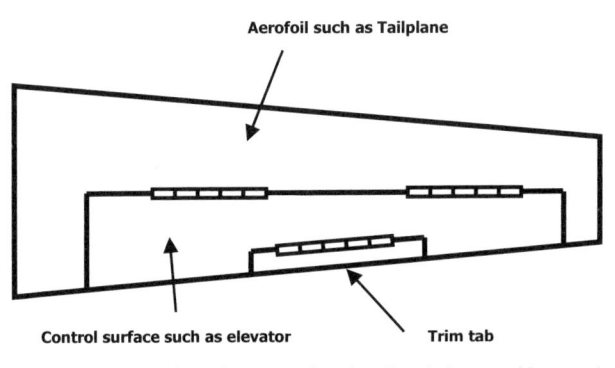

If you have ever looked closely at the elevator, aileron and rudder control surfaces of a large aircraft, you will have noticed that many have what appear to be smaller control surfaces built into them. The most common of these can be found at the back of the elevator. These are the trim tabs and are moved by the trim wheel in your aircraft. If you move the trim tab in one direction during flight (with your hands off the stick), the air pressure on the control surface changes and makes the whole assembly move in the opposite direction. So when you trim back, the tab on the elevator moves down, the air pressure moves the elevator up and this in turn pitches the plane's nose up. Remember that in normal controlled flight none of these movements actually happen because you are holding the elevator still and your trimming only changes the forces.

Not every aircraft has this level of sophistication. For example, the venerable Tiger Moth has a lever that tightens or loosens springs to assist you. At the other extreme, some aircraft such as Concorde can trim by moving fuel around to minimise the drag that would be caused by deflected control surfaces.

Why is Trimming Necessary?

The forces that affect an aircraft are normally simplified into four components. They are drag, thrust, weight and (most importantly for flight) lift. Now think of a modern jet airliner in flight. The engines are under the wing and much lower than everything else. It should be clear that increasing the power generated by these low-slung engines would force the aircraft nose higher. Conversely, reducing power must have the opposite effect and lower the nose. Even single engine propeller aircraft behave this way. In most planes the centre of gravity is in front of the centre of lift for reasons of stability but this means that they are always trying to push the nose down. In normal flight these forces pushing the nose up and down all roughly equal out but not quite. The final balancing of these forces is done with the elevator trim. However, if you change power, speed, flap setting, weight (as you burn fuel) or even if your passengers walk down the aisle, you will have to re-trim to rebalance the forces. This also explains why trimming in roll and yaw is less important. The forces involved in changes to propeller wash (affects yaw) and fuel distribution (affects roll) are far less than those from engines, flaps and lift.

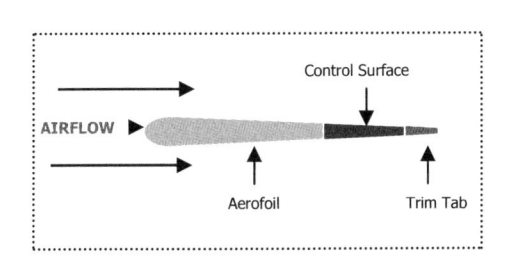

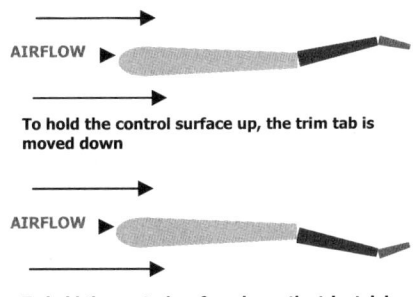

Effect of trim tab position on a control surface

Trimming Explained

When to Trim

There are two phrases that should be engraved in every pilot's brain. They are "Power Attitude Trim" and "Attitude Power Trim". The first is used when you want to start a climb, start a descent, or level off from a descent. The second is used when you want to level off from a climb. They are the order in which you change things but the key is that you sort everything else out and get the plane stable before you trim the plane.

Any change that affects drag, thrust, weight or lift will need to be trimmed out. This includes changes in power, flap, undercarriage, spoilers, fuel cross feeds and rate of climb. In fact the trim has to be changed in so many circumstances that it is an absolutely fundamental part of flying.

As you gain experience, you will find that there are times when it is sensible to add some trim early to reduce the forces before the plane has stabilised, however this is frowned on in certain circles.

When you enter a turn, or even fly aerobatics, it may seem sensible to re-trim but these changes are usually so short lived that it is not worth the hassle of re-trimming. It is also very useful to know that the plane will be in trim when you finish the turn or manoeuvre. If you are expecting to orbit for a considerable time, you may choose to re-trim, but this is the exception.

Why Pilots Have Difficulty Trimming

There are three main reasons pilots have difficulty trimming their aircraft. The first is that they do not realise that the final trim adjustments have to be tiny and slow. The pitch trim wheel in a light aircraft is a disk with ridges on it every 10 mm or so. The final trim adjustments would be about half a notch or less. This is on a wheel that takes about ten full pulls of about 100 mm to go from one extreme to the other. Scale this to a simulator and the final adjustment will be less than one pixel, in other words you probably won't see the change on screen. You will only be able to detect the change in the aircraft response. So, golden rule number one is to trim slowly.

The second reason why pilots have difficulty is that they try to trim before the plane has stabilised. The result is often confusion about which way to trim because the changes in power, speed and attitude all take place at different times. The effects of these changes are not necessarily in the same direction. What happens is that the plane seems trimmed but as it settles down, the balance changes and the plane's attitude changes, slowing or speeding up as a consequence, which changes the balance yet again. This ends up with the pilot chasing the trim, repeatedly trimming forwards and then backwards. So, golden rule number two is to let the aircraft stabilise before you trim.

The final reason why pilots have difficulty is because they do not fly the plane at all times during the trimming process. They get close to the trim point, relax on the controls and then try to do the final trim using the trim wheel alone. This results in exactly the same chasing of the trim that happens if you try to trim too soon. So the ultimate golden rule is to always fly the aeroplane with the proper controls.

Trimming - Final Points

The trimming of any aircraft, whether light plane, airliner or helicopter is essentially the same. Once you have learnt the skill on one aircraft, you can apply it to all of the others.

If you ever take any flying tests, be warned that examiners sometimes wind the trim all over the place before giving you control again to check your ability to recover from unusual attitudes. This makes recovery an order of magnitude more difficult because the feedback from the joystick gives you misleading information. It is particularly disconcerting when you are being tested IFR and limited panel at the same time.

Trimming Explained

Accurate trimming is a skill that needs to be learnt at the same time as you learn to use the joystick. It is that important to flying properly. Take any flying test and if you cannot, or do not, trim, you would fail. Reaching for the trim should become automatic and you must get to the stage where you can trim without even thinking about what you are doing. So take the time to become an expert in trimming, because until you are totally at home with it, you cannot truly be a pilot.

By Stephen Heyworth
For TecPilot

SECTION eleven

The Big Flight Challenge

The Big Flight Challenge

The big challenge this time is split into three parts and has quite a different flavour to the last book. The first part of the challenge is relatively straightforward, comprising a skip in a light single engine aeroplane from Cambridge to Norwich, a mere 50 miles that should be easily achievable by pilots just beginning to fly cross country. Once there, you will need to change planes to a twin for a radio navigation flight across the Channel to Le Touquet, which is also known as Paris Plage. You will find the radio navigation takes a little more skill and understanding but should be well within the capability of intermediate pilots. At Le Touquet you will change aircraft yet again to complete the final stage of your flight, which is a low level VFR (Visual) flight across France, through mountain valleys of Switzerland to your final landing aerodrome in Locarno. If you think that visual flying was only for the beginners, think again, flying at high speed and low level in poor visibility is for the experts. It will definitely tax your skills. The aeroplane for the third part of the challenge will be the Vought Corsair and you will be cruising at 270 knots. If you don't have this aircraft in your version of MS Flight Simulator, you will need to find an aircraft with similar performance. At this speed you will be covering four and a half miles per minute, which is a mile every thirteen seconds. In fast planes, not only can you get to your destination quickly but you can also get very lost, very quickly.

The recommended planes for the three sections are the **Cessna 182**, the **Beech Baron 58 twin** and as already mentioned the beautiful **Corsair Warbird**. The weather for all of the flights is the same. The wind is 250/30, which means that the wind is 30 knots from 250 True. The visibility is only 5 miles, which puts it at the limit for a basic PPL qualified pilot. If you have read and understood the chapter about navigation in winds, you will be able to modify these plans to suit any other plane or weather conditions. You should use the radio to talk to Air Traffic Control throughout your flight for the airport clearances and flight following.

Route **Cambridge - Norwich**

| Brakes off | | Land | |
| Take off | | Brakes on | |

Waypoint	Dist nmi	Alt feet	Track True	Wind True / kts	Vel kts	Hdg True	Drift	Var deg	Hdg Mag	TAS kts	IAS kts	GS kts	Time mins	ETA	ATA
Cambridge	37	2500	061	250 / 30		059	R02	3 W	062	126	120	156	14		
Old Buckenham	14	2500	038	250 / 30		031	R07	3 W	034	126	120	151	5		
Norwich															

Cambridge to Norwich

This flight is similar to a route that a pilot might fly as part of a qualifying cross country to gain a PPL, although they would normally fly it in a slightly slower plane such as a Piper Tomahawk. There are two legs to this flight. There is an intermediate waypoint at Old Buckenham, which has a short asphalt runway and a grass strip, although the grass strip is not visible. If you fly accurately, it is relatively easy to spot. When you start your flight from the parked position on the apron, you should write down the time you start to taxi. This is because the times that are entered into a pilot's logbook are the "brakes to brakes" times. Don't forget to ask for and receive the necessary taxi clearance and, when ready, your take off clearance. As your wheels leave the ground, write down this time too. After you have taken off, you have a choice; you can either turn at 500 feet directly onto your magnetic heading for Old Buckenham, or take the more cautious approach and "set heading" overhead the airfield. In the latter option, which is commonly used by trainee pilots, you would manoeuvre to be at your cruising altitude, speed and magnetic heading directly overhead the airport. You will need to turn through 180 degrees after take off for your first leg, so you may as well set heading overhead. Whichever method you choose, you should note the time that you start the leg. Immediately calculate your Estimated Time of Arrival (ETA) for Old Buckenham and write it down. The times that you log along the way should be written down as minutes only. So if your ETA is 11:34 am, you would simply write 34. You might change this for extremely long legs, but it is rarely, if ever, appropriate. It is critical now that you hold your planned speed, magnetic heading and altitude. To help you on your way, you should see Mildenhall just to your left after about 4 minutes, passing abeam after a further minute. One or two minutes later, you may just

The Big Flight Challenge

spot Lakenheath in the distance again to your left. After about twelve minutes, start looking for Old Buckenham. Don't just look ahead, look around everywhere, including to the sides. You would be amazed how easy it is not to see an aerodrome until you are passing it. When you finally spot Old Buckenham, fly directly to it and as you fly over it turn onto your new magnetic heading. Write down your Actual Time of Arrival (ATA) and then calculate and write down your ETA for Norwich. Two minutes later you will almost fly over the crossed runways of Hethel and soon afterwards Norwich should pop into view. You will see the City of Norwich to your right, so give the aerodrome a call for your landing instructions. Be aware that landing in this strong wind will make the groundspeed on your final approach much slower than usual, so be careful that you don't drop under the glide slope. Hopefully this leg was not too difficult. Everything rests on holding your magnetic heading, height and speed accurately and then looking for the waypoints shortly before your planned ETA.

Route **Norwich - Le Touquet**

Brakes off		Land													Set heading	
Take off		Brakes on														

Waypoint	Dist nmi	Alt feet	Track True	Wind True	/	Vel kts	Hdg True	Drift	Var deg		Hdg Mag	TAS kts	IAS kts	GS kts	Time mins	ETA	ATA
Norwich																	
	50	6000	186	250	/	30	193	L08	3	W	196	202	180	187	16		
CLN 114.55																	
	23	6000	225	250	/	30	229	L04	3	W	232	202	180	174	8		
SND 362.5																	
	17	6000	193	250	/	30	200	L07	3	W	203	202	180	184	5		
DET 117.30																	
	21	5500	150	250	/	30	158	L08	3	W	161	200	180	203	6		
LYD 114.05																	
	41	5500	136	250	/	30	144	L08	3	W	147	200	180	210	12		
Le Touquet																	

Norwich to La Touquet

After you have landed and parked, change to the Beech Baron 58 Twin for the next leg to La Touquet. This will be a visual flight but you will use navigation aids for guidance along the route. If you know how to use the radios, this should be an easy flight but it will give you an opportunity to brush up on your beacon tracking skills. Alternatively, you can use the autopilot (historically known as George for some strange reason) for this flight and that is how you are guided below. Don't worry, you will have plenty of scope for hand flying in the last part of the challenge.

If you look at the plan before you start, you will see that the frequency of each radio aid is given. A VOR's frequency always starts with a "one" and is of the form 111.75, so it is easily distinguishable from an NDB frequency, which is always a higher number such as 345 or 345.5. Hence the planned route uses three VOR's and one NDB. Good airmanship includes keeping ahead of the plane, therefore you should do as much as possible before you even release the brakes for taxi. You can tune the first VOR and NDB frequencies. Put the frequency for the Clacton VOR into Nav 1 and Nav 2 and activate them. You are unlikely to pick up the beacons while you are still on the ground. Next put the frequency for the Detling VOR into both standby windows. Now move on to the ADF and set the frequency for Southend. Set the Magnetic Track to the Clacton VOR, which is the Track True plus the Variation West making it 189 (186 plus 3 West), on the HSI Course Deviation Indicator (CDI) course arrow. Set the Heading bug on the HSI to your planned initial magnetic heading. Finally move to the Autopilot and set the initial cruising altitude of 6000 feet. With clearances in place, you can taxi, take off and set heading for the Clacton VOR. You have planned to fly at a significant altitude, so it would not be sensible to climb to your cruising altitude and then set heading over the airport. So start your timing when you pass abeam the airport in your climb-out or when you turn onto your track, whichever is most appropriate. As you climb, and turn onto your initial planned heading, activate the autopilot heading and altitude modes. Soon the VOR's will become active but do not turn to intercept the planned radial until you

The Good Flight Simmer's Guide Mk.II - 2003

The Big Flight Challenge

have identified both the beacons and DME, otherwise, who knows where you may end up. When you are happy with the Morse identification codes, change the heading bug on the HSI to converge slowly with and then track the planned VOR radial. Only when you are stable on the VOR, change to the autopilot's navigation mode.

Whilst you have been doing all of this, the radio will doubtless have been chattering to you. As a pilot you need to handle this at the same time as your flying, it's all part of your job. However, you should only have any conversation when you have the plane fully under control and heading in (or turning to) the correct direction.

The DME to Clacton will start to wind down from about 50 miles but once you are flying on the autopilot's navigation mode you can set the heading bug (not the CDI course arrow) to the heading that you want to turn on after Clacton. As you are flying in navigation mode on the autopilot, this will have no effect. The reason you have done it now is so that when you pass over the Charlie Lima November beacon, you will be able to switch the autopilot to heading mode with one press of a button, making the plane instantly turn onto the new heading. If you are busy with the radio, noting the time, or something else, you should be able to squeeze this single button press in easily. So when you reach CLN, press that button and log the time as the ATA. Next turn the HSI course arrow onto the magnetic track planned to take you to the Southend NDB, again using the true track plus the variation. Use the heading bug to intercept the radial. When you are half way to SND, adjust your heading to fly directly to, and over, the beacon. As you are now tracking using the NDB, you no longer need the VOR, so switch NAV 1 to DET, ident it and set the course arrow to the magnetic track planned from SND to DET. Set the DME to show your distance from NAV 2, which is still set on CLN so that you will have a good guide for your distance to run to Southend. At the NDB you will still be in autopilot heading mode, so a simple change to the heading bug will move you onto your next magnetic heading. Note your ATA and calculate the next ETA as usual. You can now intercept the track to the Detling VOR. You should track towards DET using the autopilot in navigation mode, leaving you to preset the heading bug for the leg to Lydd. The procedure at DET is as before, simply move the autopilot into heading mode and then adjust the bug to properly intercept the VOR radial before changing the autopilot back to navigation mode. Don't forget to descend to 5500 feet for this leg. The next change is at LYD and the procedure is exactly the same as before, juggling between the autopilot's heading and navigation mode.

There is an NDB neatly placed at Le Touquet on frequency 358 and with ident Lima Tango, so set this on the ADF to guide you in. As you are travelling at about three miles per minute, you will need to start your descent well in advance, otherwise you will still be far too high when you reach your destination. You are flying at 5500 feet and the Le Touquet circuit height is 1000 feet above airport level. The airport is 36 feet above mean sea level, so the circuits are at 1036 feet, however 1000 feet is close enough for descent planning purposes. You need to descend 4500 feet and you are flying at about 180 knots, which is 3 miles per minute. If you descend at 500 feet per minute, you will need to descend for 9 minutes. You should throttle back to keep your speed constant, so at three miles per minute, your descent will take 27 miles. As there are only 41 miles between Lydd and Le Touquet, you will need to start your 500 feet per minute descent 14 miles DME from Lydd. Listen to the Le Touquet ATIS and then contact their ATC for landing instructions. ATC are likely to give you the longest runway rather than the "into wind" one, consequently you will probably have to land with a significant crosswind. As you fly around the circuit, think where the wind is blowing from and how this will affect your circuit, approach and landing.

Strictly speaking, you should approach Le Touquet from the Visual Reference Point (VRP) to the North of the zone. If you wish to do this, modify your last leg and track directly towards the BNE VOR on 113.8 (when you have idented it). The VRP is 10 miles from BNE on the 290 radial. So to get to it, fly directly towards BNE fly until you are DME 35 from LYD, then turn onto heading 210. When you cross the 290 radial from BNE, you are at the VRP, which is on the river about 4 miles south east of Boulogne. Flying southward and staying about two miles inland from the coast will then take you to Le Touquet. All of this requires some nifty juggling of the VOR radials but it should be well within your reach if you have already flown this far.

After landing, taxi to the parking ramp and refuel ready for the next sector. You are going to need lots of fuel for the 500 mile high speed journey to Lucarno.

The Big Flight Challenge

Route: **Le Touquet - Locarno**

	Brakes off			Land							Set heading		
	Take off			Brakes on									

Waypoint	Dist nmi	Alt feet	Track True	Wind True / Vel kts	Hdg True	Drift	Var deg	Hdg Mag	TAS kts	IAS kts	GS kts	Time mins	ETA	ATA
Le Touquet	24	400	161	250 / 30	167	L06	3 W	170	272	270	270	5		
Abbeville	27	400	127	250 / 30	132	L05	3 W	135	272	270	287	6		
Glisy	25	500	091	250 / 30	093	L02	2 W	095	273	270	301	5		
Peronne	33	700	162	250 / 30	169	L06	2 W	171	274	270	271	7		
Courmelles	43	700	154	250 / 30	160	L06	2 W	162	274	270	275	9		
St Remey	13	500	180	250 / 30	186	L06	2 W	188	273	270	261	3		
Romilly	15	600	137	250 / 30	143	L06	2 W	145	273	270	284	3		
Barberey	36	1100	142	250 / 30	148	L06	1 W	149	276	270	284	8		
Chatillion	30	1600	076	250 / 30	077	L01	1 W	078	279	270	308	6		
Langres	35	1600	157	250 / 30	164	L06	1 W	165	279	270	278	7		
Saint Adrien	21	1000	117	250 / 30	121	L05	1 W	122	275	270	295	4		
Thise	4	1500	182	250 / 30	188	L06	1 W	189	278	270	265	1		
La Veze	20	2900	150	250 / 30	156	L06	1 W	157	286	270	289	4		
Pontarlier	30	2900	186	250 / 30	191	L05	1 W	192	286	270	271	7		
La Cote	17	2300	061	250 / 30	060	R01	1 W	061	282	270	312	3		
La Blecherette	11	2300	119	250 / 30	124	L05	0 W	124	282	270	301	2		
Vevey	23	1800	155	250 / 30	161	L06	0 W	161	280	270	281	5		
Martigny	12	1800	060	250 / 30	058	R01	0 W	058	280	270	309	2		
Sion	17	2300	072	250 / 30	073	L00	0 W	073	282	270	312	3		
Turtmann	5	2300	090	250 / 30	092	L02	0 W	092	282	270	310	1		
Raron	23	4700	059	250 / 30	058	R01	0 W	058	295	270	325	4		
Ulrichen	16	4700	088	250 / 30	089	L02	0 E	089	295	270	324	3		
Ambri	23	3500	160	250 / 30	166	L06	0 E	166	289	270	288	5		
Locarno														

La Touquet to Locarno

Believe it or not, you've had it pretty easy so far but things are about to get really hard. You will be flying the Vought Corsair, which is a seriously fast aircraft. You will also be flying very low, at treetop level. The visibility remains poor and the wind is still pretty strong. You will also not be using any radio aids to guide you. It will be done using your eyeballs and skill as a pilot. Consequently your flying will have to be as good as it gets. Remember that if you were flying some planes of this vintage when they were in military service, you would have needed to fly this way to survive.

The Big Flight Challenge

Start by studying the flight that has been planned for you. The route tracks from Le Touquet around Paris and then south east to the Alps. When you reach the Alps, you need to cross a few high ridges before flying along Lake Geneva (Lac Leman). Next you will fly along the bottom of several large valleys that run through the Alps, before finally landing at the north end of Lago Maggiore. The first thing you should note is that although you will cover 500 nautical miles (about 575 statute miles), it will only take you about one hour 40 minutes. There are over twenty legs, with the longest taking 9 minutes, the shortest taking 1 minute and the average taking 4.5 minutes, so you can see you are going to be busy. The hardest part however, will be navigation at 4.5 miles per minute. The meteorological visibility is only 5 miles but this is effectively reduced by flying Nap of the earth. Although the planning has used altitudes, these are purely to allow calculation of the True Air Speed. You should aim to be a maximum of 200 feet above ground level at all times unless you are climbing to cross a ridge and during the final landing phase.

Before we begin the flight, here are a few general points that should help you along the way. The Corsair's engine must be leaned at higher altitudes. This means that if you don't pull the mixture lever back as you climb over some of the high mountain passes you will lose power with disastrous results. You may find it easier to monitor your ATA's and calculate your ETA's in your head for this flight. When you are flying so low, you need to have your eyes outside the cockpit as much as possible and writing things down may distract you too much. On the other hand, if you get lost, it is always good to have things written down so that you can back track, the choice is yours.

The accuracy with which you hold your magnetic heading and speed is crucial, because these combined with your timing are your only way of finding the next waypoint. Of these three, the heading and time require the most concentration, as the speed should be sufficiently stable once you have set the engine controls correctly. The speed requires more concentration in the mountains but for most of the journey, the throttle can be set and forgotten. When you turn onto your new heading at a waypoint, make sure that you turn quickly, which means a very steep turn at the speed you are travelling. If you don't, you may overshoot your track by a mile or so. The flight has been planned as accurately as possible, however there are still several things that can give problems. Firstly, the magnetic variation used may be wrong, it steadily changes. Secondly, the DI may not read accurate to one degree. Thirdly, the world is not a perfect sphere. Fourthly, the wind may not be exactly as planned, especially as you will be changing altitude as you follow the terrain up and down. These mean that there are potential differences between the calculated heading and the best one to take you along your course. A pilot would assess these as the flight progresses. We reckoned that adding one degree to each planned heading gave a better track between the waypoints. Don't just add one degree, however, assess what works for you, we may have been consistently out when reading our Direction Indicator.

Be very careful when flying down amongst hilly terrain, as the sloping hillsides can fool you into inadvertent bank attitudes. This is because your brain latches on to any ground that looks flat as being the level horizon, which it is often not. There is plenty of fuel in the Corsair, however if you fly something else, make sure that you manage your fuel properly and stop along the route if you have to. Almost all of the turning points are airports and this has been done to ensure that there is an identifiable object at the start and end of each leg. The next point is that although you are flying low and very much hands on, you must not forget to keep the aircraft trimmed at all times. This will enable you to do the other things that are vital to your flight, such as completing your flight log, monitoring your engine, handling the flight following ATC and so on. Finally, remember that with low visibility and high speed, you will not be able to see a waypoint until you are less than a minute away. Therefore you must keep looking around, front and sides, without spending more than a few seconds looking in one direction. Look to one side for 30 seconds and you could easily miss a big waypoint.

If all this looks way too hard for you, ease back on any or all of the parameters that make it difficult. As a first concession, try climbing to 500 or 1000 feet when looking for a waypoint. If this is not enough, try raising the visibility and flying at a couple of thousand feet and this will help enormously. However, for those of you who are old hands, the prospect of taking on this hard and very challenging high speed dash, will mean that you will rate yourself on how close you can keep to the ground during the whole flight.
For the record, this flight was tested using the following settings:

- **Texture quality:** Max
- **Terrain mesh complexity:** 80
- **Terrain texture size:** high
- **Autogen density:** dense
- **Scenery complexity:** very dense

The Big Flight Challenge

The Hardest Flight Begins!

Time to go flying. Taxi for take off, calling for your clearances as normal. Just before you take off, memorise the heading and speed for the first leg. Also memorise the opposite direction (reciprocal heading), which will be your initial target. As you take off, be careful to keep the Corsair straight, both on the runway and in the initial part of the climb. Raise the undercarriage as soon as you become airborne and do not turn until you have built up a respectable flying speed. The plane is hardest to handle on the ground and at low speed, but don't let this concern you because it becomes a delight to fly at the rate you will be travelling.

Turn onto your memorised reciprocal heading and after about one minute make a hard turn back towards the airport. Over the airport, set your heading and note the time. The direction indicator is of the strip variety, and will appear to work backwards, in the same manner that a magnetic compass does. Consequently, you must be very careful to turn the correct way when making heading changes. Before you reach this or any waypoint, think whether it will be a left or right turn. Once on course, quickly write down (or memorise) your ETA for Abbeville. As you are flying fast and low, you need to watch that you don't accidentally clip a treetop when there is a slight rise in the ground. Yes, you should be that low.

Le Touquet to Abbeville is 24 miles but this will only take you five minutes. Fortunately Abbeville is a reasonable size aerodrome and should pop into view at the planned time. When you get there, you will need to turn 35 degrees left. This would take about 12 seconds in a rate one turn, which is about a mile travelled. If you delay your turn by ten seconds while you check your new heading, then execute a gentle turn, you could be over one mile off track. Hard and steep turns combined with high roll rates are the order of the day for this flight. The bigger the heading change, the harder your turn must be.

Finding Glisy is straightforward. It is just after you pass the Amiens built up area to your right. Onward to Peronne the flight is even easier, especially as a road runs straight from Glisy to Peronne. Follow the convenient line featured and it will take you within half a mile of Peronne airfield. This is your only real opportunity for the alternative type of IFR (I Follow Roads). So far so good, but don't become complacent, things are about to get much harder. It is therefore crucial that you both set your heading and calculate your ETA accurately at Peronne. Your next waypoint is Courmelles and it is almost impossible to spot unless you fly directly over it. Even then you could miss it at the (lack of) height you are flying. The key is a dual carriageway just before it and knowing that that the airfield is situated on top of a small ridge.

The countryside starts to rise and fall as you fly ever onward to St Remey. Remember to keep close to the ground, it is easy to follow a hill upward only to forget that you need to drop down again on the other side. When the ground levels out you should be looking for another hard to distinguish grass strip. It is followed by a very short leg to Romilly, which has two crossed grass runways and is simple to find because it is so close. Onward to Barberey and finally an airfield that is hard to miss, being an asphalt runway just before the town of Troyes. The run to Chatillion is hilly at first but levels out again. As with all grass strips, it somehow seems to merge with the surrounding countryside.

Langres, the next waypoint, is a grass strip aerodrome yet again. The guide for finding it is the knowledge that it is on top of a hill just after both a road and a river. You are now around the halfway point but the pressure is still on. You need to keep very focused, particularly as you will soon reach the mountains.

The leg to Saint Adrien is over some very barren and sparsely populated countryside, but at least the airport can be located without too much effort. The next waypoint is not so easy to find. The grass strip at Thise is approached by crossing a low ridge and then a dual carriageway. Thise is situated just the other side of a road at the foot of a second low ridge. When you get there, carry out a quick steep turn onto your new heading because the asphalt runway of La Veze is only one minute away over the ridge. The flight to Pontarlier is across several hills and valleys, if you are flying low enough, you will be flying below the ridges of the surrounding hills for much of the time.

The hills are now turning into mountains and the next leg will take you over two lakes and a five thousand foot high ridge before you reach the small grass strip near Lac Leman. The airfield is tricky to pick out, but it is there, right on the edge of the lake. When you turn for La Blecherette you will fly out across the lake, then across land again, once more clipping the edge of the lake before rising to the airport on the top of the hill.

The Good Flight Simmer's Guide Mk.II - 2003

The Big Flight Challenge

Unlike the real world, there's not much to see at Vevey. There is a small wide based triangular promontory of land jutting out into the lake when you reach the water's edge. Rely on your planned timing and heading to make sure you are in the right place. It's a far cry from the busy town that is actually there. Turn onto your next heading across the lake to Martigny. You will enter a long wide valley running South South East with a dual carriageway visible mostly to your left. Your timing will tell you when you reach Martigny and you will see that the main valley now turns onto East North East. Follow this valley (with its flying trees!) to the airport at Sion and then onward to Turtman, Raron and Ulrichen. The heading you have planned for these sectors is now more of a guide than something you can follow rigidly. You should be flying low along the bottom of these valleys, with the mountains rising thousands of feet on either side.

The valley twists and narrows as you reach Raron but keep following it arcing to the left towards Ulrichen. Exactly at the far threshold of Ulrichen's runway, turn hard right and climb into the first hanging valley there. Your planned headings mean even less now, as you will have to follow the valleys twisting and turning all the way to Locarno. Continue climbing up the valley and as you approach the mountain tops curve to the left over the ridge and then drop leftward into another large glaciated valley. The Corsair has plenty of power for this, but only if you are managing the engine correctly. Follow this valley (and a dual carriageway) as it bends to the right to Ambri. This is not an airfield you would choose to visit in the wrong aeroplane and you will definitely need a good landing technique if you run out of fuel here. Keep following the valley down until it ends at a junction with a new large valley running left to right. Take the right hand option. You will soon come to a further valley, also running left to right, and again turn right. If you are using the Flight Following, cancel it and call Locarno for landing. You will hopefully be given a straight in landing, so keep to the right hand side of the valley and start to reduce speed. Locarno is just before you reach the end of Lake Maggiore.

You have covered some 500 miles, and if you have succeeded in this horrendous challenge at ultra-low level, without hitting pause or peeking at the map, you can congratulate yourself heartily. There remains one task for you to complete before you can finally let your guard down. Landing the Vought Corsair is not easy, especially if you have all the realism settings set to maximum. After that long journey, you will want to concentrate particularly hard because no flight is any good without a safe landing.

Most of this flight was totally illegal. Primarily because we flew at such low levels, effectively buzzing aerodromes and towns across France and Switzerland. ATC usually get upset at this kind of flying, despite it being so much fun. Then that's the beauty of flight simulation, you can do it with impunity. Now that you know the route and have probably noticed some intermediate features that may help your navigation, you should find a second run much easier than your first. So why not go back and fly it all again but even lower this time!

By Stephen Heyworth
For TecPilot

The Big Flight Challenge

Big Flight Challenge ICAO Codes

Cambridge	EGSC	Norwich	EGSH	Le Touquet	LFAT
Old Buckenham	EGSV	CLN 114.55	VOR	Abbeville	LFOI
Norwich	EGSH	SND 362.5	NDB	Glisy	LFAY
		DET 117.30	VOR	Peronne	LFAG
		LYD 114.05	VOR	Courmelles	LFJS
		Le Touquet	LFAT	St Remey	LFFZ
				Romilly	LFQR
				Barberey	LFQB
				Chatillion	LFQH
				Langres	LFSU
				Saint Adrien	LFEV
				Thise	LFSA
				La Veze	LFQM
				Pontarlier	LFSP
				La Cote	LSGP
				La Blecherette	LSGL
				Vevey	Town
				Martigny	Town
				Sion	LSGS
				Turtmann	LSMJ
				Raron	LSMN
				Ulrichen	LSMC
				Ambri	LSPM
				Locarno	LSZL

NOTES

SECTION twelve
ADVENTURES

a-z
FS Add-On
Adventures

A NOTE FROM MIKE CLARK (EDITOR)

The diverse, internationally produced products in this section are designed to give you a perspective of add-on products available for Microsoft Flight Simulator 98/2000/2002 and possibly *FS2004. All are republished, re-packaged, redeveloped, renamed, prodded, tweaked and dropped at some point in time. This is inevitable and beyond our control. Therefore, DO NOT expect them to be "available" or "as seen" when researching them further. The products featured were current at the time of publication and every effort has been made to ensure the information provided is accurate. They will be updated again in our next (year) book in 2004.

For your convenience we have included ALL developer or publisher web sites so that you can discover updated information online. Please be aware that some web sites DO NOT speak fluent English but may be affiliated with places that do (e.g. www.simMarket.com). Further, to ensure that as many products were incorporated as possible (Over 350 - 100 more than last year) we speeded up the process of typesetting by simplifying the translation process. Some product descriptions may reflect this and we apologise if you find an occasional grammatical error or "simplified sentence".

After the release of our first book we were interrogated by one magazine editor who demanded an explanation as to why we bothered including this section at all. The reasons were and still are simple; there are many thousands, possibly millions of people who don't know what products are available to download OR buy in the shops. Some are novices while others have difficulty using the Internet - I know I sometimes do. Also, many, MANY developers (and publishers) like to keep a "paper" record of what's been manufactured in the past and don't want an office cluttered up with hundreds of Flight Sim magazines. It helps them prevent repetitive or boring software releases, copyright infringement and other issues. A real pilot, developer and loyal friend recently said, "I LOVE your add-ons section. I regularly flick through it to see if there is a niche I've overlooked or who is stealing my ideas.. It also reminds me of, "All The Worlds Aircraft" - but for FS. Great Stuff!"

Abbreviations	
FS98	Flight Simulator 98
FS2000	Flight Simulator 2000
FS2002	Flight Simulator 2002
FS2004	Flight Simulator "A Century Of Flight"

* An asterisk next to an abbreviation signifies that the product MAY be compatible. However the software was not specifically designed for it. We suggest you check with developer's web sites for the compatibility issues.

Ratings	
Forget It	✦ ✦ ✦ ✦
Below Average	✦ ✦ ✦ ✦
Average	✦ ✦ ✦ ✦
Above Average	✦ ✦ ✦ ✦
Excellent!	✦ ✦ ✦ ✦

The Good Flight Simmer's Guide Mk.II - 2003

FS Add-On - Adventures   a-Z

Title: **Adventures Unlimited - Alitalia**
TecPilot Rating: ✦ ✦ ✦ ✦
Description: Developed with approval of Alitalia Company and in collaboration with VirtuAlitalia, the largest virtual airline group. The package includes: Boeing 767-300, 747-400, 777-300, A320.200, McDonnell Douglas MD-11 and MD-80, all in Alitalia livery with highly photorealistic textures and panels. Includes unlimited adventures and Alitalia Flight Manager.
Pub/Dev/Source: Perfect Flight 2000 Project/Marco Martini and others
Rec Spec: Pentium III-1GHz or higher, 256 Mb or more RAM, 32 Mb 3D graphics card.
Platform: FS2002 *FS2004
Type: Download
Web Site: http://www.fs2000.org

Title: **Adventures Unlimited - Azzurra Air**
TecPilot Rating: ✦ ✦ ✦ ✦ ✦
Description: Includes 3 aircraft in Azzurra Air livery: Boeing 737-700, Avro Regional Jet 70; Avro Regional Jet 85. Design yourself all the adventures you want using the exclusive Adventure Compiler. Includes a host of other features.
Pub/Dev/Source: Perfect Flight 2000 Project/Marco Martini and others
Rec Spec: Pentium III-1GHz or higher, 256 Mb or more RAM, 32 Mb 3D graphics card.
Platform: FS2002 *FS2004
Type: Download
Web Site: http://www.fs2000.org

Title: **Adventures Unlimited - British Airways**
TecPilot Rating: ✦ ✦ ✦ ✦ ✦
Description: Includes ABL Adventure Compiler, British Airways Flight Manager and Bae 146-300, De Havilland Dash 8-300, Jetstream 41, Bae Concorde, Boeing 747-400, Boeing 777-300, Boeing 757-200, Airbus 320-200 aircraft, Checklist Book and British Airways User Guide (in pdf format).
Pub/Dev/Source: Perfect Flight 2000 Project/Marco Martini and others
Rec Spec: Pentium III-1GHz or higher, 256 Mb or more RAM, 32 Mb 3D graphics card.
Platform: FS2002 *FS2004
Type: Download
Web Site: http://www.fs2000.org

Title: **Adventures Unlimited - Lufthansa**
TecPilot Rating: ✦ ✦ ✦ ✦
Description: Includes Bobmardier CRJ 100, Avro RJ85, Boeing 737-400, Boeing 747-400, Airbus A320.200 and Airbus A300-600 aircraft, all in Lufthansa livery. Design yourself all the adventures you want using the Adventure Compiler. Complete Lufthansa Scheduled Flight Plans to use with the AI/ATC. Lufthansa Manager. Boeing 737-700 Flight Manual and more.
Pub/Dev/Source: Perfect Flight 2000 Project/Marco Martini and others
Rec Spec: Pentium III-1GHz or higher, 256 Mb or more RAM, 32 Mb 3D graphics card.
Platform: FS2002 *FS2004
Type: Download
Web Site: http://www.fs2000.org

a-Z   FS Add-On - Adventures

Title:	**Adventures Unlimited - Northwest**
TecPilot Rating:	✦ ✦ ✦ ✦ ✦
Description:	Includes Boeing 727-200, 747-400, Boeing 757-200, Airbus A320.200, McDonnel Douglas DC10 and MD-82, all in Northwest livery. Design yourself all the adventures you want using the exclusive Adventure Compiler. Northwest Flight Manager, a software interface written in MS Access. Boeing 727-200 Flight Manual and more.
Pub/Dev/Source:	Perfect Flight 2000 Project/Marco Martini and others
Rec Spec:	Pentium III-1GHz or higher, 256 Mb or more RAM, 32 Mb 3D graphics card.
Platform:	FS2002 *FS2004
Type:	Download
Web Site:	http://www.fs2000.org

Title:	**Airline Flights 2000**
TecPilot Rating:	✦ ✦ ✦ ✦ ✦
Description:	The voices of professionals were used in these adventures. The flights begin on the docking bridge and all procedures, beginning with engine start must be completed. As an extra bonus 5 aircraft are included (3 x B747, 2 x B767).
Pub/Dev/Source:	Aerosoft
Rec Spec:	Pentium III-1GHz or higher, 256 Mb or more RAM, 32 Mb 3D graphics card.
Platform:	FS98 FS2000
Type:	Box
Web Site:	http://www.aerosoft.com

Title:	**Captain Speaking 2002**
TecPilot Rating:	✦ ✦ ✦ ✦ ✦
Description:	Captain Speaking allows you to take 37 different airline flights, accompanied by authentic air traffic control. More the 22,000 voice recordings are included. In 3 different languages. The program even makes you change squawk and altitude changes when flying to a different continent. The program also equips you with a GPSW, with MAYDAY instructions.
Pub/Dev/Source:	JustFlight
Rec Spec:	Pentium III-1GHz or higher, 256 Mb or more RAM, 32 Mb 3D graphics card.
Platform:	FS2002
Type:	Box
Web Site:	http://www.justflight.com

Title:	**Captain Speaking**
TecPilot Rating:	✦ ✦ ✦ ✦ ✦
Description:	Captain Speaking allows you to take 37 different airline flights, accompanied by authentic air traffic control. More the 22,000 voice recordings were included. In 3 different languages. The program even makes you change squawk and altitude changes when flying to a different continent. The program also equips you with a GPSW, with MAYDAY instructions.
Pub/Dev/Source:	JustFlight
Rec Spec:	350 Mhz Processor, 64Mb RAM, 290Mb HD space.
Platform:	FS2000
Type:	Box
Web Site:	http://www.justflight.com

FS Add-On - Adventures   a-Z

ADVENTURES

Title: **FS2000 ATC**
TecPilot Rating: ✈✈✈✈✈
Description: Powerful adventure creator that also adds audio ATC to several portions of the flight in-
Pub/Dev/Source: Stephen Coleman
Rec Spec: Pentium III-1GHz or higher, 256 Mb or more RAM, 32 Mb 3D graphics card.
Platform: FS98 FS2000
Type: Download
Web Site: http://www.stephencoleman.com/fs2000.html

Title: **FlightDeck Companion**
TecPilot Rating: ✈✈✈✈✈
Description: Popular add-on from Dave March, author of the award winning freeware program "S-Combo", takes flight simulation to a totally new flight-level. It adds audio enhancements to your virtual cockpit making your time on the flight deck as busy as any real pilot! Adds support for a full crew compliment, 15 different voice sets, Copilot, Captain, GPWS, Vref, gear and flap calls are provided automatically and Cabin announcements etc. etc.
Pub/Dev/Source: OnCourse Software/Dave March
Rec Spec: Pentium III-1GHz or higher, 256 Mb or more RAM, 32 Mb 3D graphics card.
Platform: FS20002 FS-ACOF
Type: Download
Web Site: http://www.oncourse-software.co.uk

Title: **FlightSim Commander**
TecPilot Rating: ✈✈✈✈✈
Description: FlightSim Commander combines all important functions for navigation when flying airliners under IFR. The program includes: Tracking of AI traffic including tail number, flight level and ground speed. Three different types of professional flight planning including SIDs and STARs. Great Circle navigation. Departure and approach routes for all airports in FS. Completely autopilot-controlled flights from take-off to landing. Display of all airports with runways, taxiways, aprons, and frequencies. Printable professional approach charts. Weather generator with up to 6 weather changes per flight. Professional weather report with wind, cloud and visibility information. Plus loads more!
Pub/Dev/Source: Aerosoft
Rec Spec: Pentium III-1GHz or higher, 256 Mb or more RAM, 32 Mb 3D graphics card.
Platform: FS2000 FS2002
Type: Box
Web Site: http://www.aerosoft.com

Title: **Fly The Mad Dog**
TecPilot Rating: ✈✈✈✈✈
Description: Includes MD-83 in 25 liveries and a panel designed exclusively for this package. As you would expect from Perfect Flight 2000 a large set of Adventures written in ABL and thrown in for good measure.
Pub/Dev/Source: Perfect Flight 2000 Project/Marco Martini and others
Rec Spec: Pentium III-1GHz or higher, 256 Mb or more RAM, 32 Mb 3D graphics card.
Platform: FS2002 *FS2004
Type: Download
Web Site: http://www.fs2000.org

a-Z   FS Add-On - Adventures

Title: General Aviation
TecPilot Rating: ✦✦✦✦✦
Description: General Aviation allows you to request different flight levels during a flight and generates background ATC conversations. You can select and set V1 and VR speeds and the Decision Height. Varied weather conditions are generated individually during each flight. Flights include plans and start situations (set to engine off).
Pub/Dev/Source: Perfect Flight 2000 Project/Marco Martini and others
Rec Spec: Pentium III-1GHz or higher, 256 Mb or more RAM, 32 Mb 3D graphics card.
Platform: FS2002 *FS2004
Type: Download
Web Site: http://www.fs2000.org

Title: Jumbo 2003
TecPilot Rating: ✦✦✦✦✦
Description: This package includes eighteen planes, ABL Adventure Compiler, Jumbo Manager and extensive documentation. Features a B747-400 flight manual.
Pub/Dev/Source: Perfect Flight 2000 Project/Marco Martini and others
Rec Spec: Pentium III-1GHz or higher, 256 Mb or more RAM, 32 Mb 3D graphics card.
Platform: FS2002 *FS2004
Type: Download
Web Site: http://www.fs2000.org

Title: Radar Contact Version 3
TecPilot Rating: ✦✦✦✦✦
Description: Provides one of the most authentic ATC environments available today. This version, builds upon the success of versions 1 and 2, which were adventure generators. Due to the inherent limitations of the adventure language version 3 has been totally re-written in Visual Basic and runs as a stand-alone program while Flight Simulator is running. Radar Contact Version 3 has over 60 ATC functions not available in FS2002 ATC. These functions cover everything from multiple flight planner support, airport departures, en-route procedures, arrival airports, terminal and non terminal airport support. Also the user interface is totally revamped and the virtual co-pilot is greatly enhanced.
Pub/Dev/Source: JDT LLC
Rec Spec: Pentium III-1GHz or higher, 256 Mb or more RAM, 32 Mb 3D graphics card.
Platform: FS98 FS2000 FS2002 *FS2004
Type: Download or Box
Web Site: http://www.jdtllc.com

Title: Real ATC/Three Real Flights
TecPilot Rating: ✦✦✦✦✦
Description: Three Flight Simulator adventures with nearly 7 hrs of real flight time. Recorded from actual cockpits during real flights. Over 22 hours of actual recordings. Single keystroke settings for weather changes, radio frequencies and more. Includes original RealATC & RealATC2 (free) and Instruction Manual.
Pub/Dev/Source: REALATC/GSG Entertainment/Ralph Zimmermann
Rec Spec: Pentium III-1GHz or higher, 256 Mb or more RAM, 32 Mb 3D graphics card.
Platform: FS2000 FS2002 *FS2004
Type: Download
Web Site: http://www.realatc.com

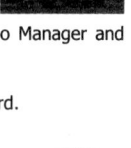

FS Add-On - Adventures   a-Z

ADVENTURES

Title: **Triple Seven**
TecPilot Rating: ★★★★★
Description: Package includes twenty-eight aircraft; 431 AI/ATC Adventures; 75 Perfect Flight 2000 Adventures, B777 sounds and the Triple Seven Manager. A software package that allows you the completely manage your flights!
Pub/Dev/Source: Perfect Flight 2000 Project/Marco Martini and others
Rec Spec: Pentium III-1GHz or higher, 256 Mb or more RAM, 32 Mb 3D graphics card.
Platform: FS2002 *FS2004
Type: Download
Web Site: http://www.fs2000.org

Title: **United Airlines Flight Ops**
TecPilot Rating: ★★★★★
Description: Allows you to request different flight levels during the flight, generates all the background ATC conversations. You can select and set the V1 and VR speeds and the Decision Height. Fully functioning, all clearance (start-up engines, taxi. line-up, take-off, climb, descend etc.), GPWS (Ground Proximity Warning System), check lists, auto take-off, SID and STAR and complete routing. Flights are rated and a score given. Automatic flight duration and fuel consumption reports - and loads more!
Pub/Dev/Source: Perfect Flight 2000 Project/Marco Martini and others
Rec Spec: Pentium III-1GHz or higher, 256 Mb or more RAM, 32 Mb 3D graphics card.
Platform: FS2002 *FS2004
Type: Download
Web Site: http://www.fs2000.org

Our online databases contain thousands of SIDS and STAR charts,
Airport Diagrams, Tutorials, News, Reviews and MORE!

WWW.TECPILOT.COM
Home Of "The Good Flight Simmer's Guide"

a-Z
FS Add-On
Aircraft

FS Add-On - Aircraft   a-Z

Title: **737-500 for Fly!**
TecPilot Rating: ✦✦✦✦✦
Description: Wilco remains one of the few companies who ventured into the world of Fly! add-ons. The 737 package includes an ultra high resolution, detailed panel, as well as 14 different aircraft liveries. The exterior leaves something to be desired, the panel raises an eye-
Pub/Dev/Source: Wilco Publishing
Rec Spec: Pentium III-1GHz or higher, 256 Mb or more RAM, 32 Mb 3D graphics card.
Platform: Fly!
Type: Box
Web Site: http://www.wilcopub.com

Title: **767 Pilot in Command**
TecPilot Rating: ✦✦✦✦✦
Description: The 767 Pilot in Command package puts you in some of the most terrifying situations that an airline pilot could ever experience. You'll challenge fires, depressurisations, and other problems. The plane and panels included are recreated in top quality.
Pub/Dev/Source: Wilco Publishing
Rec Spec: Pentium III-1GHz or higher, 256 Mb or more RAM, 32 Mb 3D graphics card.
Platform: FS2000 FS2002
Type: Box
Web Site: http://www.wilcopub.com

Title: **A Century of Aviation - Lucky Lindbergh**
TecPilot Rating: ✦✦✦✦✦
Description: Lucky Lindbergh is the first in a line of products designed to coincide with the celebration of 100 years of powered flight. Contains three aircraft that were special to Lindbergh: The Jenny, The Spirit of St. Louis and The Lockheed Sirius. Completely designed in GMAX, all can be flown from the virtual cockpit. Scenery is included for take off and landing of the Atlantic crossing. Historic documentation provided.
Pub/Dev/Source: Lago
Rec Spec: Pentium III-1GHz or higher, 256 Mb or more RAM, 32 Mb 3D graphics card.
Platform: FS2002 *FS2004
Type: Download
Web Site: http://www.lagoonline.com

Title: **A-10 Thunderbolt II**
TecPilot Rating: ✦✦✦✦✦
Description: This A-10 Thunderbolt comes with a very detailed exterior model, as well as a 3D, look around panel. The A-10, or "warthog" as it is affectionately known, was designed for FS002 by Dave Eckert. This A-10 can be downloaded and used for 7 days before registering is required.
Pub/Dev/Source: Abacus
Rec Spec: Pentium III-1GHz or higher, 256 Mb or more RAM, 32 Mb 3D graphics card.
Platform: CFS2 FS2000 FS2002
Type: Download
Web Site: http://www.flightsimdownloads.com

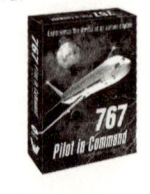

a-Z   FS Add-On - Aircraft

Title:	**A320 - Pilot In Command**
TecPilot Rating:	✦✦✦✦✦
Description:	From the company that brought you 767 Pilot in Command, comes a ground-breaking sequel based one of Europe's top airliners.
Pub/Dev/Source:	Wilco Publishing
Rec Spec:	Pentium III-1GHz or higher, 256 Mb or more RAM, 32 Mb 3D graphics card.
Platform:	FS2002 *FS2004
Type:	Box
Web Site:	http://www.wilcopub.com

Title:	**A320 - Professional**
TecPilot Rating:	✦✦✦✦✦
Description:	Phoenix takes flight again with their incredible A320 Professional package for Microsoft Flight Simulator 2002. Just Flight's boxed edition includes 35 liveries and a A320 cockpit video.
Pub/Dev/Source:	Just Flight/Phoenix Simulations
Rec Spec:	Pentium III-1GHz or higher, 256 Mb or more RAM, 32 Mb 3D graphics card.
Platform:	FS2002 *FS2004
Type:	Box
Web Site:	http://www.phoenix-simulation.co.uk

Title:	**Abacus DC-3**
TecPilot Rating:	✦✦✦✦✦
Description:	The DC-3 is considered to be the "classic" airliner of all time. For years it was the workhorse of many commercial airline companies after having served in World War II as the C-47. Luckily for us, there are still many DC-3s flying today. NOW FREE!
Pub/Dev/Source:	Abacus
Rec Spec:	Pentium III-1GHz or higher, 256 Mb or more RAM, 32 Mb 3D graphics card.
Platform:	FS98 FS2000
Type:	Download
Web Site:	http://www.flightsimdownloads.com

Title:	**Adventures Unlimited - Alitalia**
TecPilot Rating:	✦✦✦✦✦
Description:	Developed with approval of Alitalia Company and in collaboration with VirtuAlitalia, the largest virtual airline real related. The package includes: Boeing 767-300, 747-400, 777-300, A320.200, McDonnell Douglas MD-11 and MD-80, all in Alitalia livery with highly photorealistic textures and panels. Unlimited Adventures and the famous The Alitalia Flight Manager.
Pub/Dev/Source:	Perfect Flight 2000 Project/Marco Martini and others
Rec Spec:	Pentium III-1GHz or higher, 256 Mb or more RAM, 32 Mb 3D graphics card.
Platform:	FS2002 *FS2004
Type:	Download
Web Site:	http://www.fs2000.org

The Good Flight Simmer's Guide Mk.II - 2003

FS Add-On - Aircraft   a-Z

Title: **Adventures Unlimited - Azzurra Air**
TecPilot Rating: ✦✦✦✦✦
Description: Includes 3 aircraft in Azzurra Air livery: Boeing 737-700, Avro Regional Jet 70; Avro Regional Jet 85. Design yourself all the adventures you want using the exclusive Adventure Compiler. Includes a host of other features.
Pub/Dev/Source: Perfect Flight 2000 Project/Marco Martini and others
Rec Spec: Pentium III-1GHz or higher, 256 Mb or more RAM, 32 Mb 3D graphics card.
Platform: FS2002 *FS2004
Type: Download
Web Site: http://www.fs2000.org

Title: **Adventures Unlimited - British Airways**
TecPilot Rating: ✦✦✦✦✦
Description: Includes: ABL Adventure Compiler, British Airways Flight Manager and Bae 146-300, De Havilland Dash 8-300, Jetstream 41, Bae Concorde, Boeing 747-400, Boeing 777-300, Boeing 757-200, Airbus 320-200 aircraft, CheckList Book and British Airways User Guide (in pdf format).
Pub/Dev/Source: Perfect Flight 2000 Project/Marco Martini and others
Rec Spec: Pentium III-1GHz or higher, 256 Mb or more RAM, 32 Mb 3D graphics card.
Platform: FS2002 *FS2004
Type: Download
Web Site: http://www.fs2000.org

Title: **Adventures Unlimited - Lufthansa**
TecPilot Rating: ✦✦✦✦✦
Description: Includes Bobmardier CRJ 100, Avro RJ85, Boeing 737-400, Boeing 747-400, Airbus A320.200, Airbus A300-600 aircraft - all in Lufthansa livery. Design yourself all the adventures you want using the Adventure Compiler. Complete Lufthansa Scheduled Flight Plans to use with the AI/ATC. Lufthansa Manager. Boeing 737-700 Flight Manual and more.
Pub/Dev/Source: Perfect Flight 2000 Project/Marco Martini and others
Rec Spec: Pentium III-1GHz or higher, 256 Mb or more RAM, 32 Mb 3D graphics card.
Platform: FS2002 *FS2004
Type: Download
Web Site: http://www.fs2000.org

Title: **Adventures Unlimited - Northwest**
TecPilot Rating: ✦✦✦✦✦
Description: Includes Boeing 727-200, 747-400, Boeing 757-200, Airbus A320.200, McDonnel Douglas DC10 and MD-82, all in Northwest livery. Design yourself all the adventures you want using the exclusive Adventure Compiler. Northwest Flight Manager, a software interface written in MS Access. Boeing 727-200 Flight Manual and more.
Pub/Dev/Source: Perfect Flight 2000 Project/Marco Martini and others
Rec Spec: Pentium III-1GHz or higher, 256 Mb or more RAM, 32 Mb 3D graphics card.
Platform: FS2002 *FS2004
Type: Download
Web Site: http://www.fs2000.org

a-Z FS Add-On - Aircraft

Title: **Aerolite 103 Ultralight**

TecPilot Rating: ✦✦✦✦✦

Description: Contains the popular and award winning Aerolite-103 ultralight from Aero-Works in wheels, skis, and floats configurations. Each of the three models has a different and authentic colour scheme for the sails or coverings.

Pub/Dev/Source: Flight Sim Models
Rec Spec: Pentium III-1GHz or higher, 256 Mb or more RAM, 32 Mb 3D graphics card.
Platform: FS2002 *FS2004
Type: Download
Web Site: http://www.flightsimmodels.com

Title: **Air Power The Cold War FS Classics Range**

TecPilot Rating: ✦✦✦✦✦

Description: Fly the might of NATO and Warsaw Pact Air Forces in this interesting expansion.
Pub/Dev/Source: Just Flight
Rec Spec: Pentium III-1GHz or higher, 256 Mb or more RAM, 32 Mb 3D graphics card.
Platform: FS98 FS2000 CFS
Type: Box
Web Site: http://www.justflight.com

Title: **Airbus Collection**

TecPilot Rating: ✦✦✦✦✦

Description: Over 50 detailed airliners with authentic panels, sounds and animations are included.
Pub/Dev/Source: Just Flight/Pilots
Rec Spec: Pentium III-1GHz or higher, 256 Mb or more RAM, 32 Mb 3D graphics card.
Platform: FS98 FS2000 CFS
Type: Box
Web Site: http://www.justflight.com

Title: **Airbus Professional**

TecPilot Rating: ✦✦✦✦✦

Description: Phoenix has taken a leap into the Airbus world, with their first Airbus A319, 320, and 321 planes. The planes are modelled and textured in Phoenix's legendary quality. Various airline liveries are available at the Phoenix website.

Pub/Dev/Source: Phoenix Simulation Software
Rec Spec: Pentium 600, 256 MB RAM, 3D Graphics Card
Platform: FS2002
Type: Payware
Web Site: http://www.phoenix-simulation.co.uk

FS Add-On - Aircraft   a-Z

Title: **Airline Pilot 1**

TecPilot Rating: ✦✦✦✦✦

Description: Airline Pilot 1 is quite possible the most complete collection of adventures for flight sim. The 6 flights all includes outstanding audio, as well as various charts. Several planes are also include in the package.

Pub/Dev/Source: Aerosoft
Rec Spec: Pentium III-1GHz or higher, 256 Mb or more RAM, 32 Mb 3D graphics card.
Platform: FS98 FS2000
Type: Box
Web Site: http://www.aerosoft.com

Title: **Airport 2000 Volume 1**

TecPilot Rating: ✦✦✦✦✦

Description: A stunning upgrade providing new aircraft, adventures and airports that bring Flight Simulator to life!

Pub/Dev/Source: Wilco Publishing
Rec Spec: Pentium III-1GHz or higher, 256 Mb or more RAM, 32 Mb 3D graphics card.
Platform: FS98 FS2000
Type: Box
Web Site: http://www.wilcopub.com

Title: **Airport 2000 Volume 2**

TecPilot Rating: ✦✦✦✦✦

Description: The sequel takes wing with seven new airports, nine new aircraft, With high fidelity Instrument Panels and 10 all-new interactive adventures.

Pub/Dev/Source: Wilco Publishing
Rec Spec: Pentium III-1GHz or higher, 256 Mb or more RAM, 32 Mb 3D graphics card.
Platform: FS98 FS2000
Type: Box
Web Site: http://www.wilcopub.com

Title: **Airport 2002 Volume 1**

TecPilot Rating: ✦✦✦✦✦

Description: Wilco return with their exclusive animation technology that brings astounding levels of realism to the world's favourite airports. Upgrade your airports to the next generation.

Pub/Dev/Source: Wilco Publishing
Rec Spec: Pentium III-1GHz or higher, 256 Mb or more RAM, 32 Mb 3D graphics card.
Platform: FS2002 *FS2004
Type: Box
Web Site: http://www.wilcopub.com

a-Z FS Add-On - Aircraft

Title: **B314 Clipper**
TecPilot Rating: ✦✦✦✦✦
Description: Boeing only built 12 of these magnificent beasts. Includes GMAX rendered aircraft in various liveries, virtual cockpit and even scenery of numerous sea bases to fly from!
Pub/Dev/Source: Just Flight
Rec Spec: Pentium III-1GHz or higher, 256 Mb or more RAM, 32 Mb 3D graphics card.
Platform: FS98 FS2000 FS2002 *FS2004
Type: Box
Web Site: http://www.justflight.com

Title: **Battle of Britain Memorial Flight**
TecPilot Rating: ✦✦✦✦✦
Description: Normally reserved for the elite of today's RAF, fly the 11 historic aircraft that make up the Battle of Britain Memorial Flight. All the planes are modelled to a high level of detail and come in the BBMF livery for 2001.
Pub/Dev/Source: Just Flight /Blue Arrow FS
Rec Spec: Pentium III-1GHz or higher, 256 Mb or more RAM, 32 Mb 3D graphics card.
Platform: CFS2 FS2000 or FS2002 *FS2004
Type: Box
Web Site: http://www.justflight.com

Title: **Beech Baron**
TecPilot Rating: ✦✦✦✦✦
Description: The Baron is one of aviation's favourite twin and Terry Hill has captured this plane with this slick rendition. It combines a fabulous visual model with full moving parts, accurate flight characteristics and a crisp, clean instrument panel.
Pub/Dev/Source: Abacus
Rec Spec: Pentium III-1GHz or higher, 256 Mb or more RAM, 32 Mb 3D graphics card.
Platform: FS2000 FS2002
Type: Download
Web Site: http://www.flightsimdownloads.com

Title: **Boeing 747**
TecPilot Rating: ✦✦✦✦✦
Description: This "try before you buy" 747 includes a very detailed exterior, as well as a high resolution instrument panel. Also included are sounds.
Pub/Dev/Source: Abacus
Rec Spec: Pentium III-1GHz or higher, 256 Mb or more RAM, 32 Mb 3D graphics card.
Platform: FS98 FS2000
Type: Download
Web Site: http://www.flightsimdownloads.com

FS Add-On - Aircraft   a-Z

Title: **Boeing 757/767 For Fly!**
TecPilot Rating: ✦✦✦✦✦
Description: The package includes 12 757s and 767s, painted in various airline schemes. The planes have been included with some of the most accurate flight dynamics and panels ever for Fly!
Pub/Dev/Source: Aerosoft
Rec Spec: Pentium 266 minimum (recommended), 64 Mb Ram, Microsoft Flight Simulator 98, CD
Platform: Fly! Fly!2k
Type: Box
Web Site: http://www.aerosoft.com

Title: **Boeing 777-200 Professional**
TecPilot Rating: ✦✦✦✦✦
Description: 777-200 is quite possible the most accurately designed aircraft ever. The body is extremely detailed, as well is the advanced panel. The sounds for 777 PRO are all digitally recorded.
Pub/Dev/Source: JustFlight
Rec Spec: Pentium 266 minimum (recommended), 64 Mb Ram, Microsoft Flight Simulator 98, CD
Platform: FS2000
Type: Box
Web Site: http://www.justflight.com

Title: **Boeing B314 - The Clipper (Online Version)**
TecPilot Rating: ✦✦✦✦✦
Description: Contains 2 Clipper Models (B314 & B314A) in standard livery. Incorporates the following Clipper bases: Auckland, Canton Island, Guam, Manila, Noumea, Oahu/Pearl Harbor, Treasure Island (San Francisco Bay), Wake Island, Bermuda, Botwood, Foynes, Horta (Azores), Shediac and Southampton. Plus the standard "LE" scenery of Foynes, Botwood, Treasure Island and Hawaii.
Pub/Dev/Source: Pilot's
Rec Spec: Pentium III-1GHz or higher, 256 Mb or more RAM, 32 Mb 3D graphics card.
Platform: FS2002 *FS2004
Type: Download
Web Site: http://www.b314clipper.com

Title: **Cessna 310**
TecPilot Rating: ✦✦✦✦✦
Description: This orange painted Cessna 310 from Abacus includes an exterior model, panel, as well
Pub/Dev/Source: Abacus
Rec Spec: Pentium III-1GHz or higher, 256 Mb or more RAM, 32 Mb 3D graphics card.
Platform: FS2000 FS2002
Type: Download
Web Site: http://www.flightsimdownloads.com

a-Z   FS Add-On - Aircraft

Title:	**Classic Airliners 2000**
	FS Classics Range
TecPilot Rating:	✦ ✦ ✦ ✦ ✦
Description:	The world's finest airliners from 1940 to 1980! Includes 200 aircraft with VIP panels and sounds. The Kings of Sky enter the new millennium with their best collection yet of aviation's early commercial airliners. Over 200 aircraft have been completely updated and enhanced for Microsoft Flight Simulator 2000.
Pub/Dev/Source:	Just Flight
Rec Spec:	Pentium III-1GHz or higher, 256 Mb or more RAM, 32 Mb 3D graphics card.
Platform:	FS98 FS2000
Type:	Box
Web Site:	http://www.justflight.com

Title:	**Classic Eipper - Formance MX**
TecPilot Rating:	✦ ✦ ✦ ✦ ✦
Description:	Contains the classic but out-of-production Eipper-Formance single-seat MX and two-seat MXII aircraft from 1982 in wheeled and float versions. These fun little planes have an authentic and simple but challenging 2-axis flight control system that flies just a little different. (Be sure to read the '2_axis_control.txt' document during your product's installation to correctly setup the two-axis flight control system.).
Pub/Dev/Source:	Flight Sim Models
Rec Spec:	Pentium III-1GHz or higher, 256 Mb or more RAM, 32 Mb 3D graphics card.
Platform:	FS2002 *FS2004
Type:	Download
Web Site:	http://www.flightsimmodels.com

Title:	**Combat Aces**
TecPilot Rating:	✦ ✦ ✦ ✦ ✦
Description:	Scenery and missions expansion for Combat Flight Simulator 2! Includes 20 aircraft, authentic WWI scenery and 70 missions spread over six campaigns.
Pub/Dev/Source:	Just Flight
Rec Spec:	Pentium III-1GHz or higher, 256 Mb or more RAM, 32 Mb 3D graphics card.
Platform:	CFS2
Type:	Box
Web Site:	http://www.justflight.com

Title:	**Combat Jet Trainer**
TecPilot Rating:	✦ ✦ ✦ ✦ ✦
Description:	There's never been a pilot in the air who could survive without training and a huge number of the world's modern 'Top Gun' fighter pilots learned their skills in the same incredible machine. The L-39 Albatros.
Pub/Dev/Source:	CaptainSim/Just Flight
Rec Spec:	Pentium III-1GHz or higher, 256 Mb or more RAM, 32 Mb 3D graphics card.
Platform:	CFS2 FS2000 FS2002 *FS2004
Type:	Box
Web Site:	http://www.justflight.com

The Good Flight Simmer's Guide Mk.II - 2003

FS Add-On - Aircraft   a-Z

Title: **Combat Squadron**
TecPilot Rating: ✈ ✈ ✈ ✈ ✈
Description: Combat Squadron includes 18 planes, designed by world renown designer, Terry Hill. Each plane includes sounds, panels, flight dynamics, as well as a exterior model. The package included-s some planes that are included in the stock CFS program
Pub/Dev/Source: Abacus
Rec Spec: Pentium III-1GHz or higher, 256 Mb or more RAM, 32 Mb 3D graphics card.
Platform: CFS1 CFS2 FS2000
Type: Box
Web Site: Http://www.abacuspub.com

Title: **Concorde SST**
TecPilot Rating: ✈ ✈ ✈ ✈ ✈
Description: This Concorde is like no other for FS98. It has a very detailed exterior, as well as a very complete panel. The plane is available in a British Airway livery.
Pub/Dev/Source: Abacus
Rec Spec: Pentium III-1GHz or higher, 256 Mb or more RAM, 32 Mb 3D graphics card.
Platform: FS98 FS2000 FS2002
Type: Download
Web Site: http://www.flightsimdownloads.com

Title: **Corporate Pilot**
TecPilot Rating: ✈ ✈ ✈ ✈ ✈
Description: Corporate Pilot painstakingly re-creates 5 of the most popular sleek and sexy biz jets in the industry. All include detailed panels, flight models, and sounds.
Pub/Dev/Source: Abacus
Rec Spec: Pentium III-1GHz or higher, 256 Mb or more RAM, 32 Mb 3D graphics card.
Platform: FS2000 FS2002
Type: Box
Web Site: http://www.abacuspub.com

Title: **DeHaviland Chipmunk**
TecPilot Rating: ✈ ✈ ✈ ✈ ✈
Description: Famous RAF low-wing, two-place, single engine aircraft with tail wheel trainer. Primarily used for training at various RAF Squadrons in the late 40s and early 50s when more than 700 aircraft were made.
Pub/Dev/Source: Abacus
Rec Spec: Pentium III-1GHz or higher, 256 Mb or more RAM, 32 Mb 3D graphics card.
Platform: FS2002
Type: Download
Web Site: http://www.flightsimdownloads.com

a-Z   FS Add-On - Aircraft

Title: Double Trouble
TecPilot Rating: ✦✦✦✦✦
Description: North American P-51D "Mustang" and Focke Wulf FW 190A8 "Butcher Bird". Rolling wheels - sliding canopies - pilot ejection - working shocks and suspension - fully rendered virtual cockpits with working gauges - multi resolution format for increased online playability and frame rates - working drop tanks - individual component separation occurs when damage potential is reached. Includes TextoMatic and the DT Expansion Pack
Pub/Dev/Source: Flight 1
Rec Spec: Pentium III-1GHz or higher, 256 Mb or more RAM, 32 Mb 3D graphics card.
Platform: CFS2 FS2002 *FS2004
Type: Download or Box
Web Site: http://www.flight1.com

Title: DreamFleet Cardinal
TecPilot Rating: ✦✦✦✦✦
Description: Modeled after N1384C, a real 1978 Cessna 177B flown by the Knoxville Flyers - a real flying club in Knoxville TN. This package contains a Virtual Cockpit & Cabin, Complete Moving Parts, Visible Interior and Passengers, Fuselage and Lighting Reflections, Accurate Flight Dynamics and more!
Pub/Dev/Source: Flight 1 / DreamFleet
Rec Spec: Pentium III-1GHz or higher, 256 Mb or more RAM, 32 Mb 3D graphics card.
Platform: FS2002 *FS2004
Type: Download
Web Site: http://www.flight1.com

Title: Eurowings Professional/ Commuter Airliners
TecPilot Rating: ✦✦✦✦✦
Description: Contains the Eurowings fleet of Aircraft: ATR-42 Version 300 and 500, ATR-72 Version 200 and 210, BAe 146 Version 200 and Airbus A319 Version 100 in various other liveries. Also contains 6 European airports: Dortmund, Headquarter of the Eurowings AG. Berlin - Tempelhof (one of the most traditional Airports of Europe). Paris - Charles-de-Gaulle. Amsterdam - Schiphol. London - Stansted. Olbia, Costa Smeralda at Sardinia.
Pub/Dev/Source: Aerosoft
Rec Spec: Pentium III-1GHz or higher, 256 Mb or more RAM, 32 Mb 3D graphics card.
Platform: FS2002 *FS2004
Type: Box
Web Site: http://www.aerosoft.com

Title: Executive Jets
TecPilot Rating: ✦✦✦✦✦
Description: Fly two of the world's most desirable business jets. The Cessna Citation X and the Hawker 800 XP.
Pub/Dev/Source: Just Flight
Rec Spec: Pentium III-1GHz or higher, 256 Mb or more RAM, 32 Mb 3D graphics card.
Platform: FS2000
Type: Box
Web Site: http://www.justflight.com

FS Add-On - Aircraft   a-Z

Title: **F-16 Thunderbird**
TecPilot Rating: ✦✦✦✦✦
Description: This is the F-16 Flighting Falcon. It is widely used for both air combat and ground attack. This one is dressed in the Thunderbird Demonstration colours but is outfitted with armaments for CFS2.
Pub/Dev/Source: Abacus
Rec Spec: Pentium III-1GHz or higher, 256 Mb or more RAM, 32 Mb 3D graphics card.
Platform: CFS2 FS2002
Type: Download
Web Site: http://www.flightsimdownloads.com

Title: **F4U-5 Corsair**
TecPilot Rating: ✦✦✦✦✦
Description: Alain L'Homme has created this striking replica of the Corsair. This unique aircraft features moving parts: rotating propeller, landing gear that retract as smooth-as-silk; flaps that drop down, sliding transparent canopy and engine cowling, moving ailerons and elevator. This Corsair is tuned for Combat Flight Simulator only and includes a damage profile.
Pub/Dev/Source: Abacus
Rec Spec: Pentium III-1GHz or higher, 256 Mb or more RAM, 32 Mb 3D graphics card.
Platform: CFS CFS2 FS2000 FS2002
Type: Download
Web Site: http://www.flightsimdownloads.com

Title: **Piper Meridian**
TecPilot Rating: ✦✦✦✦✦
Description: Built by pilots for pilots. Comes with an impressive array of avionics systems, panels and nicely constructed flight model. Redundant displays are included for both pilot and copilot. With the Meggitt avionics (MAGIC) suite, Garmin GNS 530, and S-Tec system 550 autopilot. Nice!
Pub/Dev/Source: Flight 1 / Reality-XP
Rec Spec: Pentium III-1GHz or higher, 256 Mb or more RAM, 32 Mb 3D graphics card.
Platform: FS2002 *FS2004
Type: Download or Box
Web Site: http://www.flight1.com

Title: **Flight Deck II**
TecPilot Rating: ✦✦✦✦✦
Description: Update to the original Flight Deck Blue Angels product. Superb aircraft and scenery modelling together with many effects that are still "top gun" today.
Pub/Dev/Source: Abacus
Rec Spec: Pentium III-1GHz or higher, 256 Mb or more RAM, 32 Mb 3D graphics card.
Platform: CFS2 FS2000 FS2002
Type: Box
Web Site: http://www.abacuspub.com

a-Z   FS Add-On - Aircraft

Title:	**Flightstar II SL/SC Series**
TecPilot Rating:	✈ ✈ ✈ ✈ ✈
Description:	Contains models of the sporty and fast Flightstar two seat SL and SC in wheel, skis and float configurations. Each of the four models has a different and authentic colour scheme for the sails or coverings.
Pub/Dev/Source:	Flight Sim Models
Rec Spec:	Pentium III-1GHz or higher, 256 Mb or more RAM, 32 Mb 3D graphics card.
Platform:	FS2002 *FS2004
Type:	Download
Web Site:	http://www.flightsimmodels.com

Title:	**Fly The Best**
TecPilot Rating:	✈ ✈ ✈ ✈ ✈
Description:	Fly The Best, comes with 18 modern airliners. They all have detailed flight and visual models, and instrument panels, all in different airline paint schemes. The bonus software of Instant Airplane maker, allows you to create your own paint schemes and more!
Pub/Dev/Source:	Abacus
Rec Spec:	Pentium III-1GHz or higher, 256 Mb or more RAM, 32 Mb 3D graphics card.
Platform:	FS98 FS2000
Type:	Box
Web Site:	http://www.abacuspub.com

Title:	**Fly The Mad Dog**
TecPilot Rating:	✈ ✈ ✈ ✈ ✈
Description:	Includes Md-83 in 25 liveries and a panel designed exclusively for this package. As you would expect from Perfect Flight 2000 a large set of Adventures written in ABL and thrown in to-boot.
Pub/Dev/Source:	Perfect Flight 2000 Project/Marco Martini and others
Rec Spec:	Pentium III-1GHz or higher, 256 Mb or more RAM, 32 Mb 3D graphics card.
Platform:	FS2002 *FS2004
Type:	Download
Web Site:	http://www.fs2000.org

Title:	**Fly to Hawaii: DC10**
TecPilot Rating:	✈ ✈ ✈ ✈ ✈
Description:	The Hawaii Airline DC10 was meticulously recreated by FlightSoft for the Fly to Hawaii add-on. The plane's biggest feature is it's round "wide body" fuselage. New techniques were used to give the plane an almost round look.
Pub/Dev/Source:	FlightSoft
Rec Spec:	Microsoft Flight Simulator 2000/98 Pentium II 300 PC, Windows 98, 64 Mb Ram, CD ROM Drive, Sound Card.
Platform:	FS2000 FS2002
Type:	Box
Web Site:	http://www.flightsoft.com

FS Add-On - Aircraft   a-Z

Title: **GeeBee Racer**
TecPilot Rating: ✦✦✦✦✦
Description: Graphic artist Scott Nix recreates this spectacular reproduction the famous GeeBee, a craft from a bygone era. The paint scheme is unlike any that you're ever seen before. This truly amazing plane is made using FS Design Studio. Includes Panel.
Pub/Dev/Source: Abacus
Rec Spec: Pentium III-1GHz or higher, 256 Mb or more RAM, 32 Mb 3D graphics card.
Platform: FS2000 FS2002
Type: Download
Web Site: http://www.flightsimdownloads.com

Title: **General Aviation Collection 421C Golden Eagle**
TecPilot Rating: ✦✦✦✦✦
Description: The 421C is one of the most highly detailed aircraft available for Flight Simulator 2002 and is designed in muti-resolution format. The Golden Eagle steps up to the plate and defines what detail is all about. We dare say that you will be hard pressed to find another FS model with this level of detail and accuracy.
Pub/Dev/Source: Flight 1
Rec Spec: Pentium III-1GHz or higher, 256 Mb or more RAM, 32 Mb 3D graphics card.
Platform: FS2002 *FS2004
Type: Box
Web Site: http://www.flight1.com

Title: **Glassair III**
TecPilot Rating: ✦✦✦✦✦
Description: The Glassair is a high performance kit plane. For those who'd rather not spend the time to build the kit, Terry Hill has assembled it for you using the FS Design Studio tool. This cool plane can easily reach 200 knots and fly 1200+ miles.
Pub/Dev/Source: Abacus
Rec Spec: Pentium III-1GHz or higher, 256 Mb or more RAM, 32 Mb 3D graphics card.
Platform: FS2000 FS2002
Type: Download
Web Site: http://www.flightsimdownloads.com

Title: **Greatest Airliners 737-400**
TecPilot Rating: ✦✦✦✦✦
Description: One of the most accurate recreations of an aircraft ever created for flight simulation. Designed to let you experience and learn about the most popular commercial jet airliner ever made, almost nothing has been overlooked.
Pub/Dev/Source: Flight One/Dreamfleet 2000/JustFlight
Rec Spec: Pentium III-1GHz or higher, 256 Mb or more RAM, 32 Mb 3D graphics card.
Platform: FS2000 FS2002 *FS2004
Type: Box
Web Site: http://www.flight1.com

a-Z   FS Add-On - Aircraft

Title:	**Greatest Airplanes: Archer!**
TecPilot Rating:	✦✦✦✦✦
Description:	This is not just any Archer but perhaps the most famous Archer in the flight simulation world: N8439T, a 1982 Archer II and the personal aircraft of Nels Anderson, web master of FlightSim.com, the world's largest flight simulation web site!.
Pub/Dev/Source:	Just Flight/Dreamfleet/Flight One
Rec Spec:	Pentium III-1GHz or higher, 256 Mb or more RAM, 32 Mb 3D graphics card.
Platform:	FS2002 *FS2004
Type:	Box
Web Site:	http://www.flight1.com

Title:	**Guardians of the Sky**
TecPilot Rating:	✦✦✦✦✦
Description:	Three sophisticated surveillance aircraft to fly. Avro Shackleton, BAe Nimrod MR.2 and Boeing E-3D Sentry.
Pub/Dev/Source:	Just Flight
Rec Spec:	Pentium III-1GHz or higher, 256 Mb or more RAM, 32 Mb 3D graphics card.
Platform:	FS2000
Type:	Box
Web Site:	http://www.justflight.com

Title:	**Harrier Jets**
TecPilot Rating:	✦✦✦✦✦
Description:	Vertical take off and formidable fire power. Graham Oxtoby and Dennis Seeley bring you 7 models: the AV-8B II, AV-8B II Plus, RN-FA2, GR Mk7, GR Mk7 XX, FRS2 and GR Mk3.
Pub/Dev/Source:	Abacus
Rec Spec:	Pentium III-1GHz or higher, 256 Mb or more RAM, 32 Mb 3D graphics card.
Platform:	FS2002
Type:	Download
Web Site:	http://www.flightsimdownloads.com

Title:	**Harrier Jump Jet 2002**
TecPilot Rating:	✦✦✦✦✦
Description:	Includes 23 Aircraft, Falklands Scenery and Combat Missions. The military's most remarkable aircraft is now available to fly. This comprehensive expansion provides a detailed model of the legendary 'Jump Jet'.
Pub/Dev/Source:	Just Flight
Rec Spec:	Pentium III-1GHz or higher, 256 Mb or more RAM, 32 Mb 3D graphics card.
Platform:	CFS CFS2 FS2000 FS2002 *FS2004
Type:	Box
Web Site:	http://www.justflight.com

FS Add-On - Aircraft   a-Z

Title: **Iron Knuckles DC-9-30**
TecPilot Rating: ✦✦✦✦✦
Description: The cockpit allows you to access multiple sub-panel assemblies for a proper pilots perspective. The DC-9 includes high quality sounds and Text-o-Matic aircraft repainting utility. Text-o-Matic allows you to download and install additional aircraft paint schemes. Includes documentation, checklists, specifications and aircraft history.
Pub/Dev/Source: Flight 1
Rec Spec: Pentium III-1GHz or higher, 256 Mb or more RAM, 32 Mb 3D graphics card.
Platform: FS2002 *FS2004
Type: Download or Box
Web Site: http://www.flight1.com

Title: **KC135-R Stratotanker**
TecPilot Rating: ✦✦✦✦✦
Description: When the military aircraft need mid-air refuelling, it calls on the KC-135. Based on the design of the venerable Boeing 707, this tanker can feed a litter with 200,000 pounds of fuel. Flies at 500 mph over a range of 1600 miles. By Rey Lopez.
Pub/Dev/Source: Abacus
Rec Spec: Pentium III-1GHz or higher, 256 Mb or more RAM, 32 Mb 3D graphics card.
Platform: FS2002
Type: Download
Web Site: http://www.flightsimdownloads.com

Title: **Korean Combat Pilot**
TecPilot Rating: ✦✦✦✦✦
Description: North and South take to the skies as jets duel with propellers in CFS. Includes 14 Aircraft, Korean Scenery, 20 Combat Missions and Campaigns.
Pub/Dev/Source: Just Flight
Rec Spec: Pentium III-1GHz or higher, 256 Mb or more RAM, 32 Mb 3D graphics card.
Platform: CFS CFS2
Type: Box
Web Site: http://www.justflight.com

Title: **L-1011 Tristar FS Classics Range**
TecPilot Rating: ✦✦✦✦✦
Description: 16 airline liveries of the L-1011 -100, 200, 250. Plus bonus liveries for free download. Fully animated with moving gear, spoilers, flaps, ailerons, rudder, elevators and fan blades.
Pub/Dev/Source: Just Flight
Rec Spec: Pentium III-1GHz or higher, 256 Mb or more RAM, 32 Mb 3D graphics card.
Platform: FS98 FS2000
Type: Box
Web Site: http://www.justflight.com

a-Z   FS Add-On - Aircraft

Title: **Legendary MiG-21 Special**
TecPilot Rating: ✦✦✦✦✦
Description: Includes three MiG-21UM Versions: Standard (one 490L tank and 2 APU-13 rails), Clean (2 APU-13 rails) and Long Range (one 800L tank and 2x490L tanks) in 10 versions which include: USSR, USAF, RAF, Finnish AF, GDR AF, Czechoslovakian AF, Polish AF, Indian AF and Luftwaffe.
Pub/Dev/Source: Flight 1
Rec Spec: Pentium III-1GHz or higher, 256 Mb or more RAM, 32 Mb 3D graphics card.
Platform: FS2002 *FS2004
Type: Download
Web Site: http://www.flight1.com

Title: **Luftwaffe Collection**
TecPilot Rating: ✦✦✦✦✦
Description: Provides a representative selection of the Luftwaffe's finest from 1918 to present day. The Zeppelin to the EF2000, World War 1 and 2 fighters and bombers, Cold War strategic aircraft, there's even a Mig 29 in Unified German colours plus some interesting historical aircraft such as the twin rotor Flettner helicopter!
Pub/Dev/Source: Just Flight
Rec Spec: Pentium III-1GHz or higher, 256 Mb or more RAM, 32 Mb 3D graphics card.
Platform: CFS FS98 FS2000
Type: Box
Web Site: http://www.justflight.com

Title: **M-Squared Breese 2 Series**
TecPilot Rating: ✦✦✦✦✦
Description: Contains M-Squared Aircraft's Breese 2 series of high-quality planes of their single surface (SS) and double surface (DS) wing versions in two-place aircraft for wheels, skis, floats and amphibious configurations. Each of the six models has a different and authentic colours scheme for the sails or coverings.
Pub/Dev/Source: Flight Sim Models
Rec Spec: Pentium III-1GHz or higher, 256 Mb or more RAM, 32 Mb 3D graphics card.
Platform: FS2002 *FS2004
Type: Download
Web Site: http://www.flightsimmodels.com

Title: **M-Squared Sprint/Sport 1000 Series**
TecPilot Rating: ✦✦✦✦✦
Description: Contains the bigger and heftier M-Squared Aircraft's 1000 series of light-sport planes that are double surface (Sport) and single surface (Sprint) wing versions in two-place aircraft for wheels, skis, floats and amphibious configurations. Each of the six models has a different and authentic colour scheme for the sails or coverings.
Pub/Dev/Source: Flight Sim Models
Rec Spec: Pentium III-1GHz or higher, 256 Mb or more RAM, 32 Mb 3D graphics card.
Platform: FS2002 *FS2004
Type: Download
Web Site: http://www.flightsimmodels.com

FS Add-On - Aircraft   a-Z

Title: **MIG-21UM Special**
TecPilot Rating: ✦✦✦✦✦
Description: Simply Awesome! Includes animated adjustable afterburners, hinged canopies, instructors cockpit periscope, rotating wheels, landing gears, wheel nacelle covers, engine nozzle, all-movable stabilizer, ailerons, rudder, flaps, airbrakes, landing gears shock struts... its all to much.. we need a rest just looking at it! Comes in at a whopping three MiG-21UM versions and ten TPS versions.
Pub/Dev/Source: Captain Sim
Rec Spec: Pentium III-1GHz or higher, 256 Mb or more RAM, 32 Mb 3D graphics card.
Platform: FS2002 CFS2
Type: Download
Web Site: http://www.aerol39.com

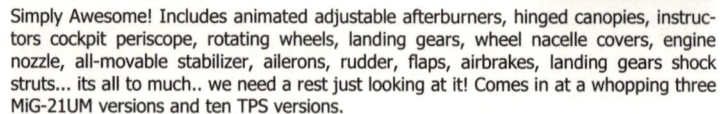

Title: **Messerschmitt 262**
TecPilot Rating: ✦✦✦✦✦
Description: The "secret weapon" that never quite made it into production towards the end of WWII. One of the first "jets" which served as an example for later aircraft. With moving parts, custom sounds and panel and ready for combat missions.
Pub/Dev/Source: Abacus
Rec Spec: Pentium III-1GHz or higher, 256 Mb or more RAM, 32 Mb 3D graphics card.
Platform: CFS1 CFS2 FS2000 FS2002
Type: Download
Web Site: http://www.flightsimdownloads.com

Title: **Mooney M20R Ovation**
TecPilot Rating: ✦✦✦✦✦
Description: This slick plane is a "personal airliner" from Mooney, a name that is synonymous with fast, high performers. The newest model Ovation cruises at up to 190 knots and will take you an incredible 1300 miles.
Pub/Dev/Source: Abacus
Rec Spec: Pentium III-1GHz or higher, 256 Mb or more RAM, 32 Mb 3D graphics card.
Platform: FS2000 FS2002
Type: Download
Web Site: http://www.flightsimdownloads.com

Title: **Mosquito Squadron**
TecPilot Rating: ✦✦✦✦✦
Description: Includes a detailed Mosquito FB. Mk VI, complete with custom designed instrument panels, gauges, flight model and sounds. You'll also find authentic scenery of RAF Methwold and the village of Methwold itself.
Pub/Dev/Source: Just Flight
Rec Spec: Pentium III-1GHz or higher, 256 Mb or more RAM, 32 Mb 3D graphics card.
Platform: CFS2 FS2000 FS2002 *FS2004
Type: Box
Web Site: http://www.justflight.com

a-z FS Add-On - Aircraft

Title:	**Mustang Vs Fw 190**
TecPilot Rating:	✦✦✦✦✦
Description:	Take control of a Mustang or Fw 190 and sit in front of an authentic panel with working custom gauges or look out over a rendered virtual panel. Features include sliding canopies, rolling wheels, moving suspension and drop tanks! New paint schemes available to download or you can use Flight One's Text-O-Matic to create your own.
Pub/Dev/Source:	Just Flight
Rec Spec:	Pentium III-1GHz or higher, 256 Mb or more RAM, 32 Mb 3D graphics card.
Platform:	CFS2 FS2002 *FS2004
Type:	Box
Web Site:	http://www.justflight.com

Title:	**Operation Barbarossa**
TecPilot Rating:	✦✦✦✦✦
Description:	Join the Luftwaffe's elite in IL2 Sturmovik from the invasion of Russia in 1941 to the gates of Berlin in 1945. Fly 230 missions in a huge campaign, or 390 in single-player mode, plus bonus missions, 64 custom skins and dogfight maps.
Pub/Dev/Source:	Just Flight
Rec Spec:	Pentium III-1GHz or higher, 256 Mb or more RAM, 32 Mb 3D graphics card.
Platform:	CFS2
Type:	Box
Web Site:	http://www.justflight.com

Title:	**P-51 Dakota Kid Mustang**
TecPilot Rating:	✦✦✦✦✦
Description:	This P-51, nicknamed "The Dakota Kid" features moving parts: rotating propeller, smoothly retracting landing gear, large drop-down flaps, a sliding canopy and more. This P-51 is ready to do battle with Combat Flight Simulator and includes a damage profile. In short it's ready for combat missions! (and also flies with FS2000)
Pub/Dev/Source:	Abacus
Rec Spec:	Pentium III-1GHz or higher, 256 Mb or more RAM, 32 Mb 3D graphics card.
Platform:	FS2000 FS2002 CFS CFS2
Type:	Download
Web Site:	http://www.flightsimdownloads.com

Title:	**P-61 Black Widow**
TecPilot Rating:	✦✦✦✦✦
Description:	Designed as the only night fighter of World War II by Northrop, the P-61 had a distinguished career against both the German Luftwaffe and Japanese IAF. With special instrumentation and realistic radial engine sounds.
Pub/Dev/Source:	Abacus
Rec Spec:	Pentium III-1GHz or higher, 256 Mb or more RAM, 32 Mb 3D graphics card.
Platform:	CFS2 FS2002 FS2000
Type:	Download
Web Site:	http://www.flightsimdownloads.com

FS Add-On - Aircraft   a-Z

Title:	**PBY-5 Catalina**
TecPilot Rating:	✦ ✦ ✦ ✦ ✦
Description:	The PBY-5 Catalina is a historic aircraft that's been used on both land and sea, in civilian and military cases. The amphibious package includes an impressive exterior, along with a detailed panel. The panel even includes moving wipers!
Pub/Dev/Source:	Abacus
Rec Spec:	Pentium III-1GHz or higher, 256 Mb or more RAM, 32 Mb 3D graphics card.
Platform:	CFS2 FS2000 FS2002
Type:	Box
Web Site:	http://www.abacuspub.com

Title:	**Pacific Combat Pilot**
TecPilot Rating:	✦ ✦ ✦ ✦ ✦
Description:	Pacific Combat Pilot is a dense package which includes 27 new planes, two new campaigns and 20 new mission. The package also puts some new scenery into CFS, such as aircraft carriers, as well as photo-realistic scenery covering many of the Pacific Island.
Pub/Dev/Source:	JustFlight
Rec Spec:	P200 MHz or higher CPU. 32MB RAM. 3D graphics accelerator card.
Platform:	CFS
Type:	Box
Web Site:	http://www.justflight.com

Title:	**Pacific Theatre**
TecPilot Rating:	✦ ✦ ✦ ✦ ✦
Description:	Pacific Theatre puts you right in the middle of WW2's hottest battles. The package also puts some new scenery into CFS, such as aircraft carriers, as well as photo-realistic elevated mesh scenery covering many of the Pacific Islands.
Pub/Dev/Source:	Abacus
Rec Spec:	Pentium III-1GHz or higher, 256 Mb or more RAM, 32 Mb 3D graphics card.
Platform:	CFS CFS2
Type:	Box
Web Site:	http://www.abacuspub.com

Title:	**Pearl Harbour**
TecPilot Rating:	✦ ✦ ✦ ✦ ✦
Description:	Relive the surprise attack that forever changed American and Japanese lives. December 7, 1941. Includes aircraft, scenery, historically accurate missions and unique 'what if' scenarios.
Pub/Dev/Source:	Flight One/Just Flight
Rec Spec:	Pentium III-1GHz or higher, 256 Mb or more RAM, 32 Mb 3D graphics card.
Platform:	CFS2
Type:	Box
Web Site:	http://www.flight1.com

a-Z FS Add-On - Aircraft

Title:	**Phoenix 757-200**
TecPilot Rating:	✦ ✦ ✦ ✦ ✦
Description:	7 different airline liveries - Condor, DutchBird, Iberia, British Airways (x2), Airtours and ATA. Multiple cockpit views - that include an operational overhead and pedestal panel. Programmable FMC - that supports FS2000 flight plans and LOTS more.
Pub/Dev/Source:	Phoenix/Just Flight
Rec Spec:	Pentium III-1GHz or higher, 256 Mb or more RAM, 32 Mb 3D graphics card.
Platform:	FS2000
Type:	Box
Web Site:	http://www.justflight.com

Title:	**Phoenix Bonanza**
TecPilot Rating:	✦ ✦ ✦ ✦ ✦
Description:	Fly three Beechcraft Bonanza aircraft (Beechcraft V35 V Tail, 33C Aerobat and A36 Turboprop) which include: Panoramic 360-degree cockpit views, Panels with custom-built gauges including a working programmable GPS. Animated gear and control surfaces - gear, gear doors, flaps, elevators, rudder, spoilers and ailerons. A lot more could be pointed out but we don't have the space.
Pub/Dev/Source:	Phoenix/Just Flight
Rec Spec:	Pentium III-1GHz or higher, 256 Mb or more RAM, 32 Mb 3D graphics card.
Platform:	FS2000 FS2002 *FS2004
Type:	Box
Web Site:	http://www.justflight.com

Title:	**Piper Aztec**
TecPilot Rating:	✦ ✦ ✦ ✦ ✦
Description:	The Piper PA-23 is more commonly known as the Aztec. This six-place twin was originally launched in the early 1960's. With two-250 hp engines, it can cruise at 205 mph at 21,000 feet and travel more than 1150 miles. Add-on maker Dave Eckert has designed this Aztec for FS2000 users with the help of FS Design Studio.
Pub/Dev/Source:	Abacus
Rec Spec:	Pentium III-1GHz or higher, 256 Mb or more RAM, 32 Mb 3D graphics card.
Platform:	FS2000 FS2002
Type:	Download
Web Site:	http://www.flightsimdownloads.com

Title:	**Piper Comanche 250**
TecPilot Rating:	✦ ✦ ✦ ✦ ✦
Description:	The popular Comanche is a retractable single that can whiz along at 155 knots with a range of 600 nautical miles. Experience flying Terry Hill's rendition of this great aircraft.
Pub/Dev/Source:	Abacus
Rec Spec:	Pentium III-1GHz or higher, 256 Mb or more RAM, 32 Mb 3D graphics card.
Platform:	FS2000 FS2002
Type:	Download
Web Site:	http://www.flightsimdownloads.com

The Good Flight Simmer's Guide Mk.II - 2003

FS Add-On - Aircraft   a-Z

AIRCRAFT

Title: **Piper Cub**
TecPilot Rating: ✦✦✦✦✦
Description: Since its introduction in the mid-1940's, the Piper Cub has become synonymous with general aviation. Designer Matt Garry is a Piper Cub owner himself so he has experienced the joy of flying this "vintage" aircraft with the original "old style" cockpit.
Pub/Dev/Source: Abacus
Rec Spec: Pentium III-1GHz or higher, 256 Mb or more RAM, 32 Mb 3D graphics card.
Platform: FS98 FS2000 FS2002 CFS
Type: Download
Web Site: http://www.flightsimdownloads.com

Title: **Piper Navajo**
TecPilot Rating: ✦✦✦✦✦
Description: Terry Hill and Jim Rhoads have teamed up to produce this beauty using their very own new cutting edge tool FS Design Studio. This six/eight place twin is a popular business plane. It can carry enough fuel for a 800 to 1000 mile trip at 215 mph. This is a "next generation" aircraft.
Pub/Dev/Source: Abacus
Rec Spec: Pentium III-1GHz or higher, 256 Mb or more RAM, 32 Mb 3D graphics card.
Platform: FS2000 FS2002
Type: Download
Web Site: http://www.flightsimdownloads.com

Title: **Piper Seneca**
TecPilot Rating: ✦✦✦✦✦
Description: One of the first aircraft additions for FS2002. Includes opening doors, reflection effects, rotating wheels among other stunning details.
Pub/Dev/Source: Phoenix Simulation
Rec Spec: 800MHz Processor, 128Mb RAM, Sound Card, 3D Accelerator Card
Platform: FS2002
Type: Download
Web Site: http://www.phoenix-simulation.co.uk

Title: **Private Pilot**
TecPilot Rating: ✦✦✦✦✦
Description: Private Pilot is a collection of 13 private propeller aircraft. All 13 planes feature moving parts (propellers, flaps, gear, etc), as well as detailed panels, flight and visual models. The program also comes with aircraft checklists, and other great additions.
Pub/Dev/Source: Abacus
Rec Spec: Pentium III-1GHz or higher, 256 Mb or more RAM, 32 Mb 3D graphics card.
Platform: FS2000
Type: Box
Web Site: http://www.abacuspub.com

a-Z   FS Add-On - Aircraft

Title: **Private Wings**
TecPilot Rating: ✈✈✈✈✈
Description: Private Wings features 36 quick and agile aircraft, each with a fully-functional cockpit and moving parts. Your first class flight is complete with original instrumentation, checklists, flight data, and radio with standby frequency. Clocks with stopwatch functionality allow for holding patterns and you can even log the number of hours in flight.
Pub/Dev/Source: Data Becker
Rec Spec: Pentium III-1GHz or higher, 256 Mb or more RAM, 32 Mb 3D graphics card.
Platform: FS2000
Type: Box
Web Site: http://www.databecker.com

Title: **Ready For Pushback**
TecPilot Rating: ✈✈✈✈✈
Description: Based on 747-200. Includes detailed fuel, hydraulic, electrical and other functional panels. You can even fuel your aircraft from the wing fuelling panel. Over 400 switches and gauges. Based on FS-2002 GMAX Models. 4 animated aircraft. Liveries include: Virgin Atlantic, Air Canada, Northwest, KLM with more available for download.
Pub/Dev/Source: Flight 1
Rec Spec: Pentium III-1GHz or higher, 256 Mb or more RAM, 32 Mb 3D graphics card.
Platform: FS2002 *FS2004
Type: Box
Web Site: http://www.flight1.com

Title: **Revolutionary EE\ BAC Lightning**
TecPilot Rating: ✈✈✈✈✈
Description: Comes in three different versions: Lite, Standard or Full. From 60 to 20 liveries are available which include authentic sound sets, highly detailed models and a fully working cockpits. Depending on which package you purchase.
Pub/Dev/Source: TLUK Aircraft Design/Daniel Dunn
Rec Spec: Pentium III-1GHz or higher, 256 Mb or more RAM, 32 Mb 3D graphics card.
Platform: FS2002 *FS2004
Type: Box
Web Site: http://www.bac-f6.com

Title: **Royal Air Force 2000 FS Classics Range**
TecPilot Rating: ✈✈✈✈✈
Description: RAF 2000 spans the RAF from pre-WW2 Hawker Biplanes to the future with the Euro fighter EF2000. 25 RAF aircraft in total, consisting of 21 types and 4 sub-variants. Each aircraft includes semi-transparent cockpit windows. All have rotating propellers and many have fully animated flight control surfaces.
Pub/Dev/Source: Just Flight
Rec Spec: Pentium III-1GHz or higher, 256 Mb or more RAM, 32 Mb 3D graphics card.
Platform: CFS FS98 FS2000
Type: Box
Web Site: http://www.justflight.com

FS Add-On - Aircraft   a-Z

AIRCRAFT

Title: **Royal Navy Aviation Collection**
TecPilot Rating: ✦✦✦✦✦
Description: The Royal Navy Aviation Collection includes 30 of Great Britain's finest planes. Aircraft such as the De Havilland Sea Venom and the F4F Wildcat have been included. The collection also comes with several CFS missions, just as "Attack on the Bismarck" and "Operation Fuller"
Pub/Dev/Source: VFR Scenery
Rec Spec: Pentium III-1GHz or higher, 256 Mb or more RAM, 32 Mb 3D graphics card.
Platform: FS98 FS2000 CFS and CFS2
Type: Download
Web Site: http://www.vfrscenery.com/

Title: **SIAI-Marchetti SF.260**
TecPilot Rating: ✦✦✦✦✦
Description: A high definition GMAX created external model featuring five unique liveries as standard, clear reflective textures and a multitude of animated parts. Includes three 2D panels, flying guide and pilot's operating reference.
Pub/Dev/Source: RealAir Simulations
Rec Spec: Pentium III-1GHz or higher, 256 Mb or more RAM, 32 Mb 3D graphics card.
Platform: FS2002 *FS2004
Type: Download
Web Site: http://www.realairsimulations.com

Title: **Sabre vs Mig**
TecPilot Rating: ✦✦✦✦✦
Description: Fly the famous aircraft of the Korean Conflict: F-86 Sabre, MiG-15, F-51, PLUS AI aircraft: B-29, F-84, F-82, TU-2, Yak-9. Each flyable aircraft is provided with a custom panel, gauges and rendered virtual cockpit. Sliding canopies, pilot ejection, jet exhaust trails, speed brakes and even rolling wheels. Includes 20 Missions over 4 campaigns.
Pub/Dev/Source: Just Flight
Rec Spec: Pentium III-1GHz or higher, 256 Mb or more RAM, 32 Mb 3D graphics card.
Platform: CFS2
Type: Box
Web Site: http://www.justflight.com

Title: **Supermarine Spitfire**
TecPilot Rating: ✦✦✦✦✦
Description: The "SPIT" was a top fighter during WWII and many consider it to be the most important aircraft of the 40s. This Mk IX, designated as the "MX-Y" has moving parts including the rotating propeller, smoothly retracting landing gear, two-part flaps and sliding canopy.
Pub/Dev/Source: Abacus
Rec Spec: Pentium III-1GHz or higher, 256 Mb or more RAM, 32 Mb 3D graphics card.
Platform: FS2000 CFS CFS2
Type: Download
Web Site: http://www.flightsimdownloads.com

a-z

FS Add-On - Aircraft

Title: **TLK-39C Pilot Training Device**
TecPilot Rating: ✦✦✦✦✦
Description: Package includes: Hinged canopy, rotating wheels, landing gears, wheel nacelle covers, elevator, ailerons, rudder, double slotted flaps, air brakes, landing gears position indicators. 3D interior of detailed cockpit with a pilot. Authentic L-39C engine (AI-25TL turbofan) and cockpit sounds. Internal and external lighting including landing gears mounted position control lights. Two periods of history of the same aircraft (L-39C #31), during and after the Cold war are represented in authentic textures.
Pub/Dev/Source: Captain Sim
Rec Spec: Pentium III-1GHz or higher, 256 Mb or more RAM, 32 Mb 3D graphics card.
Platform: FS2000 FS2002 CF2
Type: Aircraft
Web Site: http://www.aerol39.com

Title: **The 421C Golden Eagle**
TecPilot Rating: ✦✦✦✦✦
Description: Designed with GMAX the 421C fuselage and window textures provide reflective mapping. Boasts fully animated parts and features "redundant" type displays. Custom GPS. Configuration Manager and Text O Matic all included. Overall, extremely nice!
Pub/Dev/Source: Flight 1
Rec Spec: Pentium III-1GHz or higher, 256 Mb or more RAM, 32 Mb 3D graphics card.
Platform: FS2002 *FS2004
Type: Box Download or Box
Web Site: http://www.flight1.com

Title: **The Concorde Experience**
TecPilot Rating: ✦✦✦✦✦
Description: The Package includes a fairly accurate flight/visual model with animated moving parts. The instrument panel incorporates a functional Inertial Navigation System. Comprehensive documentation on flying Concorde is thrown in to enable pilots to fly using exactly the same procedures as a real Concorde captains.
Pub/Dev/Source: PJD Software
Rec Spec: Pentium III-1GHz or higher, 256 Mb or more RAM, 32 Mb 3D graphics card.
Platform: FS2000 FS2002 *FS2004
Type: Download
Web Site: http://www.pjdsoftware.co.uk

Title: **The Dam Busters**
TecPilot Rating: ✦✦✦✦✦
Description: On 16th May 1943, 19 aircraft led by Wing Commander Guy Gibson flew on Operation 'Chastise', now commonly known as the 'The Dam Busters' raid. Lancaster ED932, AJ-G was flown by Gibson and carried the ingenious bouncing bomb designed by Barnes Wallis which was used to attack dams in Germany's Ruhr valley.
Pub/Dev/Source: Just Flight
Rec Spec: Pentium III-1GHz or higher, 256 Mb or more RAM, 32 Mb 3D graphics card.
Platform: CFS2 FS2000 FS2002 *FS2004
Type: Box
Web Site: http://www.justflight.com

FS Add-On - Aircraft   a-Z

Title: **The Piper PA28 Warrior**
TecPilot Rating: ✦✦✦✦✦
Description: There are three versions of the Piper that differ in livery and registration number. The aircraft in the package include Moving parts: flaps, ailerons, rudder, elevators and prop and boast a specially designed fuselage. Transparent cockpit windows with panels are visible wiich include special night lighting effects.
Pub/Dev/Source: Pilot's/EUCOM Airlines
Rec Spec: Pentium III-1GHz or higher, 256 Mb or more RAM, 32 Mb 3D graphics card.
Platform: FS2000 FS2002 *FS2004
Type: Download
Web Site: http://www.eucomairlines.de

Title: **The Tracker - Grumman S2-E/S2-A**
TecPilot Rating: ✦✦✦✦✦
Description: Package includes Two aircraft models, Panel and Virtual Cockpit. Three set of Textures for each version; The S2-A contains: An Italian Air Force S2-A, a Royal Dutch Navy S2-A, a U.S. Navy S2-A of VS21. The S2-E model contains: a U.S. Navy S2-E of VS37, a P-16 (S2-E) of Força Aérea Brasileira,an S2-E of Royal Australian Navy. Documentation. A set of descriptive Missions/Adventures for FS2002.
Pub/Dev/Source: Perfect Flight 2000 Project/Marco Martini/Max Taccoli/Stefano Caputo
Rec Spec: Pentium III-1GHz or higher, 256 Mb or more RAM, 32 Mb 3D graphics card.
Platform: FS2002 CFS
Type: Download
Web Site: http://www.fs2000.org

Title: **The Twin Otter**
TecPilot Rating: ✦✦✦✦✦
Description: Completely designed in GMAX. 4 versions available: wheels, skis, floats & amphibian. 20 liveries painted in the colours of different operators. Designed to be flown using the virtual cockpit. All parts of the panel can be used as separate windows designed to take full advantage of multi-monitor systems. Realistic sounds.
Pub/Dev/Source: Lago
Rec Spec: Pentium III-1GHz or higher, 256 Mb or more RAM, 32 Mb 3D graphics card.
Platform: FS2002 *FS2004
Type: Download
Web Site: http://www.lagoonline.com

Title: **Tornado**
TecPilot Rating: ✦✦✦✦✦
Description: Animated cockpit canopy, pilots, control surfaces, thrust reverser and wheels. High detail 2d cockpit including animated moving ma and rear view mirrors. -Very high detailed Virtual Cockpit with working instruments. Designed in 3D studio (bigger brother of GMAX) and includes an innovative refuelling system. Has to be seen to be believed.
Pub/Dev/Source: Lago
Rec Spec: Pentium III-1GHz or higher, 256 Mb or more RAM, 32 Mb 3D graphics card.
Platform: FS2002 *FS2004 CFS3
Type: Download
Web Site: http://www.lagoonline.com

a-Z   FS Add-On - Aircraft

Title:	**Tuskegee Fighters**
TecPilot Rating:	✦ ✦ ✦ ✦ ✦
Description:	Tuskegee Fighter combines the brilliantly designed aircraft of Scott Nix, Alain L'Homme and Terry hill, with the scenery expertise of Tim Dickens and tosses in realistic missions by Chris Steele. The package was created to honour the African-American airmen.
Pub/Dev/Source:	Abacus
Rec Spec:	Pentium III-1GHz or higher, 256 Mb or more RAM, 32 Mb 3D graphics card.
Platform:	CFS1
Type:	Box
Web Site:	http://www.abacuspub.com

Title:	**Twin Bonanza**
TecPilot Rating:	✦ ✦ ✦ ✦ ✦
Description:	This cartoon-ish looking 1960s Beach includes transparent windows and moving control
Pub/Dev/Source:	Abacus
Rec Spec:	Pentium III-1GHz or higher, 256 Mb or more RAM, 32 Mb 3D graphics card.
Platform:	FS2000 FS2002
Type:	Download
Web Site:	http://www.flightsimdownloads.com

Title:	**Ultralights**
TecPilot Rating:	✦ ✦ ✦ ✦ ✦
Description:	Ultralights spans the whole range of ultra light aviation, from the paragliders to the most modern 3 axis ultralights that are fully equipped for IFR flights. Ten Ultralight aircraft (many with different color variations) included. Boasts tested flightmodels, animation and sound recordings made from real Rotax engines. Additional features include 'hand held' instruments and an "informative manual". Nice product!
Pub/Dev/Source:	Lago
Rec Spec:	Pentium III-1GHz or higher, 256 Mb or more RAM, 32 Mb 3D graphics card.
Platform:	FS2002 *FS2004
Type:	Download
Web Site:	http://www.lagoonline.com

Title:	**Vietnam Air War**
TecPilot Rating:	✦ ✦ ✦ ✦ ✦
Description:	The Vietnam conflict on your PC. This expansion comes with thirteen flyable aircraft, scenery, the USS Enterprise aircraft carrier, eight authentic airbases for CFS2 and two for FS2002. Six AI aircraft, 50 custom weapons, drop tanks, SAM missiles, 30 custom missions and three campaigns. You can even rescue the wounded from a Huey helicopter.
Pub/Dev/Source:	Just Flight
Rec Spec:	Pentium III-1GHz or higher, 256 Mb or more RAM, 32 Mb 3D graphics card.
Platform:	CFS2 FS2002 *FS2004
Type:	Box
Web Site:	http://www.justflight.com

FS Add-On - Aircraft   a-Z

AIRCRAFT

Title:	**Warbrids Extreme**
TecPilot Rating:	✦✦✦✦✦
Description:	Warbirds Extreme is a collection of today's latest and greatest military aircraft, ranging from the F-16, to the stealth B-2. Designers include Dave Eckert, Time Taylor, and Bruce Thorson.
Pub/Dev/Source:	Abacus
Rec Spec:	Pentium III-1GHz or higher, 256 Mb or more RAM, 32 Mb 3D graphics card.
Platform:	CFS2 FS2002
Type:	Payware
Web Site:	http://www.abacuspub.com

Title:	**Wings Over China**
TecPilot Rating:	✦✦✦✦✦
Description:	Wings over China is yet another one of Abacus's plane, scenery, and mission packages for CFS. This time, the scenery covers China, as well as other sectors of Southeast Asia. The planes included are a mixed bag of fighters, and bombers.
Pub/Dev/Source:	Abacus
Rec Spec:	Pentium III-1GHz or higher, 256 Mb or more RAM, 32 Mb 3D graphics card.
Platform:	CFS1
Type:	Box
Web Site:	http://www.abacuspub.com

Title:	**World Airliners 747-400 & 777-200 Pro**
TecPilot Rating:	✦✦✦✦✦
Description:	Contains both Phoenix's 747-400 and 777-200 Professional aircraft in a double-pack. Includes the liveries: Air Canada - British Airways - Cathy Pacific - KLM - Korean Airways - Lufthansa - Northwest Airlines - Qantas - South African Airways - United Airlines - Virgin Atlantic Airways. An optional package is available that includes a PAL or NTSC video. The video takes the form of a flight from London Heathrow (LHR) into one of the world's most demanding airports - Hong Kong's Kai Tak.
Pub/Dev/Source:	Phoenix Simulations/ Just Flight
Rec Spec:	Pentium III-1GHz or higher, 256 Mb or more RAM, 32 Mb 3D graphics card.
Platform:	FS2000 FS2002 *FS2004
Type:	Box
Web Site:	http://www.justflight.com

Title:	**simTECH - Fokker DR1**
TecPilot Rating:	✦✦✦✦✦
Description:	Want an aircraft where the Pilot's scarf flutters in the wind or his arms move about? Sad! Includes high resolution 2D panel with realistic wood grain effect. Glass XML gauges with bronze Bezels. Virtual Cockpit gauges operate and are accurate to the time period both in look and operation. Custom vintage 1917-19 Fokker Dr1 style sounds. 23 Liveries in Deluxe package. 1 livery in standard package.
Pub/Dev/Source:	simTECH Flight Design
Rec Spec:	Pentium III-1GHz or higher, 256 Mb or more RAM, 32 Mb 3D graphics card.
Platform:	FS2002 *FS2004
Type:	Download
Web Site:	http://www.simtechflightdesign.com

a-Z
FS Add-On
Scenery

FS Add-On - Scenery   a-Z

Title: **Aarhus (EKAH)**
TecPilot Rating: ✦✦✦✦✦
Description: This scenery covers Aarhus airport in Denmark, a regional airport, formerly Thirstrup. Includes buildings with photorealistic textures, AI compatible, dynamic and static scenery such as baggage trucks, fuel trucks, gates with transparent windows. Night lightning effects both on apron and buildings. Correct placement of taxi signs. Fully functional "wig-wags".
Pub/Dev/Source: FS Dream Factory/Cornel Grigoriu
Rec Spec: Pentium III-1GHz or higher, 256 Mb or more RAM, 32 Mb 3D graphics card.
Platform: FS2002 *FS-ACOF
Type: Download
Web Site: http://www.fsdreamfactory.com

Title: **Airport 2000 Volume 2**
TecPilot Rating: ✦✦✦✦✦
Description: The sequel takes wing with seven new airports, nine new aircraft, With high fidelity Instrument Panels, and 10 all-new interactive adventures.
Pub/Dev/Source: Wilco Publishing
Rec Spec: Pentium III-1GHz or higher, 256 Mb or more RAM, 32 Mb 3D graphics card.
Platform: FS2000 FS2002 (Patched)
Type: Box
Web Site: http://www.wilcopub.com

Title: **Airport 2000 Volume 3**
TecPilot Rating: ✦✦✦✦✦
Description: Includes 7 airports (France: Paris Orly, Denmark: Kastrup Copenhagen, Germany : Berlin Tegel, UK: London Gatwick, USA: Denver International, USA : San Francisco International, USA : Seattle Tacoma), 3 aircraft, nice sound effects and 10 adventures for beginners, intermediates and advanced pilots
Pub/Dev/Source: Wilco Publishing
Rec Spec: Pentium 600 MHz - 128 MB RAM, 3D graphics accelerator card.
Platform: FS2000 FS2002 (Patched)
Type: Box
Web Site: http://www.wilcopub.com

Title: **Airport 2002 Volume 1**
TecPilot Rating: ✦✦✦✦✦
Description: Wilco return with their exclusive animation technology that brings astounding levels of realism to the world's favourite airports. Upgrade your airports to the next generation.
Pub/Dev/Source: Wilco Publishing
Rec Spec: Pentium III-1GHz or higher, 256 Mb or more RAM, 32 Mb 3D graphics card.
Platform: FS2002 *FS-ACOF
Type: Box
Web Site: http://www.wilcopub.com

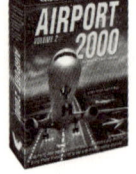

a-Z   FS Add-On - Scen-

Title: **Airport Scenery Basel 2002**
TecPilot Rating: ✥✥✥✥✥
Description: Includes highly detailed airport scenery and all major buildings. Static aircraft. New radio frequencies. ILS approach aids. Night lightning. Approach charts on CD-ROM. Switzerland 2 compatible. Dynamic Scenery. Air Inter Airbus approach on runway 16. Driving airport vehicles.
Pub/Dev/Source: Aerosoft
Rec Spec: Pentium III-1GHz or higher, 256 Mb or more RAM, 32 Mb 3D graphics card.
Platform: FS2002 *FS2004
Type: Box
Web Site: http://www.aerosoft.com

Title: **Airport Scenery Bern 2002**
TecPilot Rating: ✥✥✥✥✥
Description: Highly detailed airport of Bern-Belp. All typical airport buildings. Lighted static aircraft. All Navaids. Nice night effects. Charts on CD-Rom. Dynamic Scenery: Starting and landing aircraft. Driving buses.
Pub/Dev/Source: Aerosoft
Rec Spec: Pentium III-1GHz or higher, 256 Mb or more RAM, 32 Mb 3D graphics card.
Platform: FS2002 *FS2004
Type: Box
Web Site: http://www.aerosoft.com

Title: **Airport Scenery Lugano 2002**
TecPilot Rating: ✥✥✥✥✥
Description: Highly detailed airport of Lugano-Agno. All typical airport objects like Crossair-Hangar, Hotel La Perla, Migros city included. Static aircraft. Wonderful night textures. All NAVaids provided. Approach charts on CD-ROM and Switzerland 2 compatible plus a LOT more.
Pub/Dev/Source: Aerosoft
Rec Spec: Pentium III-1GHz or higher, 256 Mb or more RAM, 32 Mb 3D graphics card.
Platform: FS2002 *FS2004
Type: Box
Web Site: http://www.aerosoft.com

Title: **Airport Scenery Sion 2002**
TecPilot Rating: ✥✥✥✥✥
Description: Highly detailed airport of Sion. Surrounding objects like highways and castles featured. Static aircraft. All typical airport objects like restaurant and military buildings. All NAV aids. Charts on CD-ROM. Switzerland 3 compatible. Dynamics: Starting aircraft, Pilatus Porter
Pub/Dev/Source: Aerosoft
Rec Spec: Pentium III-1GHz or higher, 256 Mb or more RAM, 32 Mb 3D graphics card.
Platform: FS2002 *FS2004
Type: Box
Web Site: http://www.aerosoft.com

FS Add-On - Scenery   a-Z

Title: **Airport Scenery Zurich 2002**
TecPilot Rating: ✈ ✈ ✈ ✈ ✈
Description: A highly detailed scenery of this great airport. Zuerich-Kloten is the largest airport of Switzerland. Many features like working docking-system, dynamic scenery and much more.
Pub/Dev/Source: Aerosoft
Rec Spec: Pentium III-1GHz or higher, 256 Mb or more RAM, 32 Mb 3D graphics card.
Platform: FS2000 FS2002 *FS2004
Type: Box
Web Site: http://www.aerosoft.com

Title: **Alaska Land Class**
TecPilot Rating: ✈ ✈ ✈ ✈ ✈
Description: Covers the whole area of Alaska. Land Class works with default or third party textures. Designed from real maps, the scenery draws all geographical features such as forests, fields, autogen scenery and cities. Exactly as they exist in in real life. Fully Compatible with other Terrain Mesh Sceneries.
Pub/Dev/Source: FSFreeware/Raimondo Taburet
Rec Spec: Pentium III-1GHz or higher, 256 Mb or more RAM, 32 Mb 3D graphics card.
Platform: FS2002 *FS2004
Type: Download
Web Site: http://www.fsfreeware.com

Title: **Alicante 2002**
TecPilot Rating: ✈ ✈ ✈ ✈ ✈
Description: Features the main airport "El Altet", buildings and surrounding topography of Alicante, Spain. Includes static and dynamic objects. Photorealistic type textures and night lighting effects.
Pub/Dev/Source: SimWings/Manfred Spatz
Rec Spec: Pentium III-1GHz or higher, 256 Mb or more RAM, 32 Mb 3D graphics card.
Platform: FS2002 *FS2004
Type: Download
Web Site: http://www.sim-wings.de

Our online databases contain thousands of SIDS and STAR charts, Airport Diagrams, Tutorials, News, Reviews and MORE!

WWW.TECPILOT.COM
Home Of "The Good Flight Simmer's Guide"

a-z   FS Add-On - Scenery

Title:	**Amazonian Sceneries 1 - Belem City**
TecPilot Rating:	✦✦✦✦✦
Description:	Full photorealistic scenery of a 50 x 50 Km area. Includes Belem City and its airports. Belem is the capital of Para State - Brazil and located at the taper of Amazon River. features New Val de Caes - Intl (SBBE), Julio Cesar - Aeroclube (SBJC) and Porta do Céu - Airfield (SNPC) .
Pub/Dev/Source:	CenaReal/Jorge Luiz Padilha
Rec Spec:	Pentium III-1GHz or higher, 256 Mb or more RAM, 32 Mb 3D graphics card.
Platform:	FS2002 *FS2004
Type:	Download or Box
Web Site:	http://www.geocities.comjorgepadilha/index.html

Title:	**Amsterdam Schiphol 2002**
TecPilot Rating:	✦✦✦✦✦
Description:	Update to the original FS2000 version. Contains extensive documentation on intersections, preferential runway systems and maps that help you find correct taxi routes. The documentation contains ILS, SID and STAR information. Also features Lago's now legendary "Active Scenery" technology (AI Traffic etc).
Pub/Dev/Source:	Lago
Rec Spec:	Pentium III-1GHz or higher, 256 Mb or more RAM, 32 Mb 3D graphics card.
Platform:	FS2002 *FS2004
Type:	Download
Web Site:	http://www.lagoonline.com

Title:	**Amsterdam Schiphol Airport**
TecPilot Rating:	✦✦✦✦✦
Description:	Revamp of original FS2000 Scenery. Includes Runways and Navaids, Taxiways, roads, buildings including freight and hangars, gate facilities at all terminals, static aircraft at the piers, docking systems, AGNIS/PAPA and SAFEGATE.
Pub/Dev/Source:	MICE Innovative Computer Engineering/Hans-Jörg Müller
Rec Spec:	Pentium III-1GHz or higher, 256 Mb or more RAM, 32 Mb 3D graphics card.
Platform:	FS2000 FS2002 *FS2004
Type:	Download
Web Site:	http://www.mice.ch

Title:	**Appalachians & NE 38m Terrain**
TecPilot Rating:	✦✦✦✦✦
Description:	From Alabama and Georgia, through Kentucky, Tennessee, to West Virginia, Virginia, Maryland, Pennsylvania, New York and into all of New England, Appalachians & Northeast covers a broad area of the eastern United States. Re-sampled to 38.2m (LOD=10) from the USGS 30m National Elevation Database, detail and accuracy is ensured. .
Pub/Dev/Source:	FSGenesis
Rec Spec:	Pentium III-1GHz or higher, 256 Mb or more RAM, 32 Mb 3D graphics card.
Platform:	FS2002 *FS2004
Type:	Download or Box
Web Site:	http://www.fsgenesis.com

The Good Flight Simmer's Guide Mk.II - 2003

FS Add-On - Scenery   a-Z

Title: **Atlanta Intl (KATL)**
TecPilot Rating: ✦✦✦✦✦
Description: A detailed scenery of Hartsfield Atlanta Intl Airport (KATL) including buildings, runways, taxiways and airport objects. Realistic buildings and objects with full texturing. Active docking system and jetways at each gate. Numerous static and dynamic aircraft (parking, taxiing, taking off and landing), vehicles, objects, cars, buses, service vehicles - everything is illuminated and lit at night. Runways and taxiways including all markings and signs, with lighting. Documentation with airport map and background information.
Pub/Dev/Source: SimFlyers Associated
Rec Spec: Pentium III-1GHz or higher, 256 Mb or more RAM, 32 Mb 3D graphics card.
Platform: FS2000 FS2002 *FS2004
Type: Download
Web Site: http://www.simflyers.net

Title: **Austria Professional 2002**
TecPilot Rating: ✦✦✦✦✦
Description: Experience the mountains of Austria! This unique add-on includes the whole area of Austria with mountains, valleys, all airports and many detailed objects in cities and towns.
Pub/Dev/Source: Aerosoft
Rec Spec: Pentium III-1GHz or higher, 256 Mb or more RAM, 32 Mb 3D graphics card.
Platform: FS2000 FS2002 *FS2004
Type: Box
Web Site: http://www.aerosoft.com

Title: **Austria, Swizerland, Germany, Slovenia, Czech Republic**
TecPilot Rating: ✦✦✦✦✦
Description: A high-detail terrain mesh scenery designed at 1 Arc/Sec resolution (very high resolution). The coverage is N55 E5 to N 45 E20. Runs at high frames rates!.
Pub/Dev/Source: FSFreeware/Raimondo Taburet
Rec Spec: Pentium III-1GHz or higher, 256 Mb or more RAM, 32 Mb 3D graphics card.
Platform: FS2000 FS2002 *FS2004
Type: Box or Download
Web Site: http://www.fsfreeware.com

Title: **Austrian Airports 1-4**
TecPilot Rating: ✦✦✦✦✦
Description: This collection includes 14 of the biggest and best known airports and airfields in Austria. All airports are designed realistically and with many detailed objects. The author, Stefan Rausch, is well-known building great airport sceneries.
Pub/Dev/Source: FlightXpress/Stefan Rausch
Rec Spec: Pentium III-1GHz or higher, 256 Mb or more RAM, 32 Mb 3D graphics card.
Platform: FS2000 FS2002 *FS2004
Type: Download
Web Site: http://www.austrianairports.com

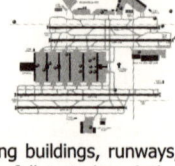

a-Z   FS Add-On - Scenery

Title: **Barcelona 2002**
TecPilot Rating: ✦✦✦✦✦
Description: Features the main airport, buildings and surrounding topography of Barcelona, Spain. Includes static and dynamic objects. Photorealistic type textures and night lighting effects.
Pub/Dev/Source: SimWings/Manfred Spatz
Rec Spec: Pentium III-1GHz or higher, 256 Mb or more RAM, 32 Mb 3D graphics card.
Platform: FS2002 *FS2004
Type: Download
Web Site: http://www.sim-wings.de

Title: **Bari Palese (LIBD)**
TecPilot Rating: ✦✦✦✦✦
Description: Includes buildings, runways, taxiways and airport objects. Realistic buildings and objects with full night and seasonal texturing. Marshaller at each parking stand. Numerous static and dynamic aircraft (parking, taxiing, taking off and landing).
Pub/Dev/Source: SimFlyers Associated
Rec Spec: Pentium III-1GHz or higher, 256 Mb or more RAM, 32 Mb 3D graphics card.
Platform: FS2000 FS2002 *FS2004
Type: Download
Web Site: http://www.simflyers.net

Title: **Berlin Tegel (EDDT)**
TecPilot Rating: ✦✦✦✦✦
Description: Detailed design and many new great looking day and night time lighting effects. Summer, winter and spring textures. Fully 32 Bit compatible. Frame rate optimized. Dynamic service and apron cars. Static aircraft with night time lighting effects and automatic changing liveries. Runways with wheel skid marks. New Terminals, runways, towers and other buildings. SAFEGATE docking system and active dock bridges.
Pub/Dev/Source: German Airports /Scenery Design Team
Rec Spec: Pentium III-1GHz or higher, 256 Mb or more RAM, 32 Mb 3D graphics card.
Platform: FS2002 *FS2004
Type: Download
Web Site: http://www.germanairports.net

Title: **Birmingham (EGBB)**
TecPilot Rating: ✦✦✦✦✦
Description: Very accurate and interactive scenery for this busy UK airport. Includes night lighting, static ground vehicles and aircraft, dynamic ground vehicles and aircraft, interactive docking air gates, food truck, working stop bars, lead off light, guard lights. All textures made to the highest standard.
Pub/Dev/Source: UK2000 Scenery/Gary Summons
Rec Spec: Pentium III-1GHz or higher, 256 Mb or more RAM, 32 Mb 3D graphics card.
Platform: FS2002 *FS2004
Type: Download
Web Site: http://www.uk2000scenery.com

The Good Flight Simmer's Guide Mk.II - 2003

FS Add-On - Scenery — a-Z

Title: **Bornholm (EKRN)**
TecPilot Rating: ✦✦✦✦✦
Description: Situated in one of the most beautiful places in Scandinavia, the island of Bornholm just between Sweden, Germany and Poland. Includes buildings with photorealistic textures, AI compatible, dynamic and static scenery such as baggage and fuel trucks. Gates with transparent windows.
Pub/Dev/Source: FS Dream Factory/Cornel Grigoriu
Rec Spec: Pentium III-1GHz or higher, 256 Mb or more RAM, 32 Mb 3D graphics card.
Platform: FS2002 *FS2004
Type: Download
Web Site: http://www.fsdreamfactory.com

Title: **Brasilia International**
TecPilot Rating: ✦✦✦✦✦
Description: Detailed "SBBR - Pres. Juscelino Kubitschek Airport". Realistic buildings and objects with full texturing. Cities most important buildings and landmarks. AI traffic compatible. Documentation with airport charts (taxi, departure, approach). Traffic files to be used with TTools reproducing real flights in the airport. Also, a Varig Log DC-10-30 Cargo aircraft, that can also be downloaded separately free from the manufacturer's website.
Pub/Dev/Source: RealFlight
Rec Spec: Pentium III-1GHz or higher, 256 Mb or more RAM, 32 Mb 3D graphics card.
Platform: FS2002 *FS2004
Type: Download
Web Site: http://www.realflight.com.br

Title: **Bromma Stockholm City Airport**
TecPilot Rating: ✦✦✦✦✦
Description: Realistically rendered airport with lots of nicely textured buildings. Photographs from the real airport were used as reference. AI compatible. Static aircraft and vehicles. Dynamic scenery. Fully functional Wig-wags (Scandinavian type). Landclass for the region and manual in PDF format.
Pub/Dev/Source: FS Dream Factory/Cornel Grigoriu
Rec Spec: Pentium III-1GHz or higher, 256 Mb or more RAM, 32 Mb 3D graphics card.
Platform: FS2002 *FS2004
Type: Download
Web Site: http://www.fsdreamfactory.com

Title: **Budapest Scenery Millennium Edition**
TecPilot Rating: ✦✦✦✦✦
Description: This extremely hi-resolution scenery of Budapest from Andras Kozma - one of flight simulation's most respected scenery designers. Package includes the city of Budapest (includes thousands of buildings) and the region's five airports. The airports include detailed control towers, terminals and taxiways. Seen to be believed, really.
Pub/Dev/Source: Andras Komez
Rec Spec: Pentium III-1GHz or higher, 256 Mb or more RAM, 32 Mb 3D graphics card.
Platform: FS2000
Type: Download
Web Site: http://www.vistamaresoft.com/

a-Z   FS Add-On - Scenery

Title: **Canada LandClass**
TecPilot Rating: ✦✦✦✦✦
Description: Covers the whole area of Canada. Land Class works with default or third party textures. Designed from real maps, the scenery draws all geographical features such as forests, fields, AutoGen scenery and cities - exactly as they are in in real life. Fully Compatible

Pub/Dev/Source: FSFreeware/Raimondo Taburet
Rec Spec: Pentium III-1GHz or higher, 256 Mb or more RAM, 32 Mb 3D graphics card.
Platform: FS2002 *FS2004
Type: Download
Web Site: http://www.fsfreeware.com

Title: **Canada**
TecPilot Rating: ✦✦✦✦✦
Description: Culmination of 4 months work. Recreating the entire country of Canada, in high-detail. A terrain mesh resolution of 1 Arc/Sec.
Pub/Dev/Source: FSFreeware/Raimondo Taburet
Rec Spec: Pentium III-1GHz or higher, 256 Mb or more RAM, 32 Mb 3D graphics card.
Platform: FS2000 FS2002 *FS2004
Type: Box
Web Site: http://www.fsfreeware.com

Title: **Central America**
TecPilot Rating: ✦✦✦✦✦
Description: Strangely named, this scenery offers complete mesh coverage of Mexico, Belize, Nicaragua, Panama, Costa Rica, Honduras, San Salvador. This is basically south central America. A demo (as with most downloadable products is available at 1.8mb covering the

Pub/Dev/Source: FSFreeware/Raimondo Taburet
Rec Spec: Pentium III-1GHz or higher, 256 Mb or more RAM, 32 Mb 3D graphics card.
Platform: FS2000 FS2002 *FS2004
Type: Box or Download
Web Site: http://www.fsfreeware.com

Title: **Channel Islands - Part 1**
TecPilot Rating: ✦✦✦✦✦
Description: Features the airports: (EGJA) Alderney, (EGJJ) Jersey, (EGJB) Guernsey. Terrain model uses specific textures for all areas (optional). All buildings accurately modelled and textured. Some Hangers doors open and close. Taxiway/stand markings that are realistic. BRITISH runway markings. Very detailed holding points with guard lights (Wig Wags). Aircraft guidance 'AGNIS' boards at EGJJ. 'Mike' the Marshaller at all airports. Unique Radars. Night lighting affects and transparent terminal glass. Airport vehicles. Perimeter fencing, walls, tress and bushes. Static and Light aircraft models. Simply great!
Pub/Dev/Source: UK2000 Scenery/Gary Summons
Rec Spec: Pentium III-1GHz or higher, 256 Mb or more RAM, 32 Mb 3D graphics card.
Platform: FS2000 FS2002 *FS2004
Type: Download
Web Site: http://www.uk2000scenery.com

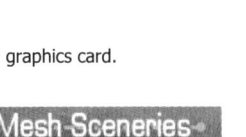

The Good Flight Simmer's Guide Mk.II - 2003

FS Add-On - Scenery   a-Z

Title: **Cologne Bonn (EDDK)**
TecPilot Rating: ✦✦✦✦✦
Description: Detailed design and many new great looking day and night time lighting effects. Summer, winter and spring textures. Fully 32 Bit compatible. Frame rate optimized. Dynamic service and apron cars. Static aircraft with night time lighting effects and automatic changing liveries. Runways with wheel skid marks. New Terminals, runways, towers and other buildings. SAFEGATE docking system and active dock bridges.
Pub/Dev/Source: German Airports /Scenery Design Team
Rec Spec: Pentium III-1GHz or higher, 256 Mb or more RAM, 32 Mb 3D graphics card.
Platform: FS2002 *FS2004
Type: Download
Web Site: http://www.germanairports.net

Title: **Dallas Fort Worth Intl (KDFW)**
TecPilot Rating: ✦✦✦✦✦
Description: Includes realistic buildings and objects with full texturing. Active docking system and jet ways. Numerous aircraft and vehicles, objects, cars, buses, service vehicles. Everything is illuminated and lit at night. Runways and taxiways including all markings and hundreds of signs, all with realistic lighting. Documentation with airport map and background information about Dallas yesterday and today. serviceArmada(TM).
Pub/Dev/Source: SimFlyers Associated
Rec Spec: Pentium III-1GHz or higher, 256 Mb or more RAM, 32 Mb 3D graphics card.
Platform: FS2000 FS2002 *FS2004
Type: Download
Web Site: http://www.simflyers.net

Title: **Dangerous Airports**
TecPilot Rating: ✦✦✦✦✦
Description: Dangerous Airports is a collection of several airports located in some of the most challenging terrain in the world! The ten airports range in location from the Aleutian Islands to Iceland. The package also includes 4 interior and exterior aircraft models.
Pub/Dev/Source: Abacus
Rec Spec: Pentium III-1GHz or higher, 256 Mb or more RAM, 32 Mb 3D graphics card.
Platform: FS98
Type: Box
Web Site: http://www.abacuspub.com

Join our online club today! We'll be happy to help!

WWW.TECPILOT.COM
Home Of "The Good Flight Simmer's Guide"

a-Z FS Add-On - Scenery

Title:	**Dortmund, Munster-Osnabruck, Paderborn-Lippstadt**
TecPilot Rating:	✦✦✦✦✦
Description:	Detailed design and many new great looking day and night time lighting effects. Summer, winter and spring textures. Fully 32 Bit compatible. Frame rate optimized. Dynamic service and apron cars. Static aircraft with night time lighting effects and automatic changing liveries. Runways with wheel skid marks.
Pub/Dev/Source:	German Airports /Scenery Design Team
Rec Spec:	Pentium III-1GHz or higher, 256 Mb or more RAM, 32 Mb 3D graphics card.
Platform:	FS2002 *FS2004
Type:	Download
Web Site:	http://www.germanairports.net

Title:	**Dresden, Monchengladback, Lubeck, Erfurt**
TecPilot Rating:	✦✦✦✦✦
Description:	Detailed design and many new great looking day and night time lighting effects. Summer, winter and spring textures. Fully 32 Bit compatible. Frame rate optimized. Dynamic service and apron cars. Static aircraft with night time lighting effects and automatic changing liveries. Runways with wheel skid marks.
Pub/Dev/Source:	German Airports /Scenery Design Team
Rec Spec:	Pentium III-1GHz or higher, 256 Mb or more RAM, 32 Mb 3D graphics card.
Platform:	FS2002 *FS2004
Type:	Download
Web Site:	http://www.germanairports.net

Title:	**Dusseldorf (EDDL)**
TecPilot Rating:	✦✦✦✦✦
Description:	Detailed design and many new great looking day and night time lighting effects. Summer, winter and spring textures. Fully 32 Bit compatible. Frame rate optimized. Dynamic service and apron cars. Static aircraft with night time lighting effects and automatic changing liveries. Runways with wheel skid marks.
Pub/Dev/Source:	German Airports /Scenery Design Team
Rec Spec:	Pentium III-1GHz or higher, 256 Mb or more RAM, 32 Mb 3D graphics card.
Platform:	FS2000 FS2002 *FS2004
Type:	Download
Web Site:	http://www.germanairports.net

Title:	**ESSB, ESMK, ESGJ & Stockholm City**
TecPilot Rating:	✦✦✦✦✦
Description:	Includes ESSB Stockholm-Bromma City Airport. Stockholm City High-Resolution photorealistic ground scenery. ESGJ Jonkoping-Axamo - a mid sized Swedish airport and ESMK Kristianstad-Everöd a smaller sized charming regional airport.
Pub/Dev/Source:	SwedFlight Pro
Rec Spec:	Pentium III-1GHz or higher, 256 Mb or more RAM, 32 Mb 3D graphics card.
Platform:	FS2002 *FS2004
Type:	Download
Web Site:	http://www.swedflight.com

FS Add-On - Scenery   a-Z

Title:	**East Anglia & North London - Part 4**
TecPilot Rating:	✦✦✦✦✦
Description:	Includes around 33 airfields in eastern England, amongst these are Heathrow and Stansted. All buildings accurately modelled and textured. Taxiway/stand markings that are realistic. BRITISH runway markings. Very detailed holding points with guard lights (Wig Wags). 3D Trees. Interactive objects. Night lighting affects and transparent terminal glass. Perimeter fencing, walls, tress and bushes. Dynamic Objects at most airfields.
Pub/Dev/Source:	UK2000 Scenery/Gary Summons
Rec Spec:	Pentium III-1GHz or higher, 256 Mb or more RAM, 32 Mb 3D graphics card.
Platform:	FS2000 FS2002 *FS2004
Type:	Download
Web Site:	http://www.uk2000scenery.com

Title:	**Eastern Europe**
TecPilot Rating:	✦✦✦✦✦
Description:	A high-detail terrain mesh scenery designed at 1 Arc/Sec resolution (very high resolution). Areas include The Balkans, Romania, Adriatic Coast, Eastern Poland, Bulgaria and more.
Pub/Dev/Source:	FSFreeware/Raimondo Taburet
Rec Spec:	Pentium III-1GHz or higher, 256 Mb or more RAM, 32 Mb 3D graphics card.
Platform:	FS2000 FS2002 *FS2004
Type:	Box or Download
Web Site:	http://www.fsfreeware.com

Title:	**Egelsbach, Bayreuth, Augsburg**
TecPilot Rating:	✦✦✦✦✦
Description:	Detailed design and many new great looking day and night time lighting effects. Summer, winter and spring textures. Fully 32 Bit compatible. Frame rate optimized. Dynamic service and apron cars. Static aircraft with night time lighting effects and automatic changing liveries. Runways with wheel skid marks. New Terminals, runways, towers and other buildings. SAFEGATE docking system and active dock bridges.
Pub/Dev/Source:	German Airports /Scenery Design Team
Rec Spec:	Pentium III-1GHz or higher, 256 Mb or more RAM, 32 Mb 3D graphics card.
Platform:	FS2000 FS2002 *FS2004
Type:	Download
Web Site:	http://www.germanairports.net

Title:	**El Hierro La Palma 2003**
TecPilot Rating:	✦✦✦✦✦
Description:	Features the main airport, buildings and surrounding topography of Palma. Includes static and dynamic objects. Photorealistic type textures and night lighting effects. A good all-rounder.
Pub/Dev/Source:	SimWings/Manfred Spatz
Rec Spec:	Pentium III-1GHz or higher, 256 Mb or more RAM, 32 Mb 3D graphics card.
Platform:	FS2002 *FS2004
Type:	Download
Web Site:	http://www.sim-wings.de

a-Z   FS Add-On - Scenery

Title:	**Emma Field**
TecPilot Rating:	✈ ✈ ✈ ✈ ✈
Description:	Far more than just scenery, Emma Field includes activities, sights and sounds of daily life at a rural airfield. From quiet winter days to busy summer weekends it's all there - in an interactive "Active Scenery" kind of way.
Pub/Dev/Source:	Lago
Rec Spec:	Pentium III-1GHz or higher, 256 Mb or more RAM, 32 Mb 3D graphics card.
Platform:	FS2002 *FS2004
Type:	Download
Web Site:	http://www.lagoonline.com

Title:	**Escalante and Hurricane**
TecPilot Rating:	✈ ✈ ✈ ✈ ✈
Description:	Escalante and Hurricane by GeoRender: Experience the wonder of Southern Utah with this highly detailed scenery of two beautiful airports. This high resolution scenery was designed using aerial photography. Includes custom made objects, ground texture made from 1 meter resolution aerial images. Sophisticated colouring to evoke the heat and light of the region. Scenery is blended in with the default terrain. Airport objects and surroundings have been created from real-life observation.
Pub/Dev/Source:	Georender
Rec Spec:	Pentium III-1GHz or higher, 256 Mb or more RAM, 32 Mb 3D graphics card.
Platform:	FS2002 *FS2004
Type:	Download
Web Site:	http://www.flight1.com

Title:	**Eurowings Professional/ Commuter Airliners**
TecPilot Rating:	✈ ✈ ✈ ✈ ✈
Description:	Contains the Eurowings fleet of Aircraft: ATR-42 300 and 500, ATR-72 200 and 210, BAe 146 Version 200 and Airbus A319 100 in various liveries. Also contains 6 European airports: Dortmund, Headquarter of Eurowings AG. Berlin - Tempelhof (one of the most traditional Airports of Europe). Paris - Charles-de-Gaulle. Amsterdam - Schiphol. London - Stansted. Olbia, Costa Smeralda at Sardinia.
Pub/Dev/Source:	Aerosoft
Rec Spec:	Pentium III-1GHz or higher, 256 Mb or more RAM, 32 Mb 3D graphics card.
Platform:	FS2002 *FS2004
Type:	Box
Web Site:	http://www.aerosoft.com

For the very latest downloads including scenery, aircraft and utilities visit...

WWW.SIMMARKET.COM
Your one stop flight shop!

FS Add-On - Scenery   a-Z

Title: **FS2002 Italy Landclass**
TecPilot Rating: ✦✦✦✦✦
Description: Covers the whole area of Italy. Land Class works with default or third party textures. Designed from real maps, the scenery draws all geographical features such as forests, fields, autogen scenery and cities. Exactly as they exist in in real life. Fully Compatible with other Terrain Mesh Sceneries.
Pub/Dev/Source: FSFreeware/Raimondo Taburet
Rec Spec: Pentium III-1GHz or higher, 256 Mb or more RAM, 32 Mb 3D graphics card.
Platform: FS2002 *FS2004
Type: Download
Web Site: http://www.fsfreeware.com

Title: **Los Angeles Area 10M Mesh & Landclass**
TecPilot Rating: ✦✦✦✦✦
Description: Covers the whole area of Los Angeles. Land Class works with default or third party textures. Designed from real maps, the scenery draws all geographical features such as forests, fields, AutoGen scenery and cities. Exactly as they exist in in real life. Fully Compatible with other Terrain Mesh Sceneries.
Pub/Dev/Source: FSFreeware/Raimondo Taburet
Rec Spec: Pentium III-1GHz or higher, 256 Mb or more RAM, 32 Mb 3D graphics card.
Platform: FS2002 *FS2004
Type: Download
Web Site: http://www.fsfreeware.com

Title: **Spain & Portugal Land Class**
TecPilot Rating: ✦✦✦✦✦
Description: Covers the whole area of Spain & Portugal. Land Class works with default or third party textures. Designed from real maps, the scenery draws all geographical features such as forests, fields, AutoGen scenery and cities. Exactly as they exist in in real life.
Pub/Dev/Source: FSFreeware/Raimondo Taburet
Rec Spec: Pentium III-1GHz or higher, 256 Mb or more RAM, 32 Mb 3D graphics card.
Platform: FS2002 *FS2004
Type: Download
Web Site: http://www.fsfreeware.com

Title: **FSLandClass**
(Scenery and Developer Packs)
TecPilot Rating: ✦✦✦✦✦
Description: A huge set of global/regional ready-made Scenery Packs (SP - Used directly within FS) and Developer Packs (DP - Ready-made scenery/data used in conjunction with a utility called FSLandClass) for users and coders to enhance Flight Simulator's default geographic/topographic environment. An impressive array of 68 areas are available (at the time of writing).
Pub/Dev/Source: Burkhard Renk
Rec Spec: Pentium III-1GHz or higher, 256 Mb or more RAM, 32 Mb 3D graphics card.
Platform: FS2002 *FS2004
Type: Download
Web Site: http://www.simmarket.com

The Good Flight Simmer's Guide Mk.II - 2003

a-Z

FS Add-On - Scenery

Title:	**Fernando de Noronha**
TecPilot Rating:	✦✦✦✦✦
Description:	Detailed airport built using Gmax™. Includes AFCAD™ files, Photorealistic terrain, Auto-Gen, Enhanced AI traffic, 3D objects of the monuments and buildings. High resolution mesh altitude.
Pub/Dev/Source:	RealFlight
Rec Spec:	Pentium III-1GHz or higher, 256 Mb or more RAM, 32 Mb 3D graphics card.
Platform:	FS2002 *FS2004
Type:	Download
Web Site:	http://www.realflight.com.br

Title:	**Tampa Intl Airport (KTPA)**
TecPilot Rating:	✦✦✦✦✦
Description:	Highly realistic virtual representation of the Tampa Intl Airport (KTPA). See below. Built with GMAX© and SCASM© for optimized performance. Custom runways and taxiways including all taxi signs and night lighting. 6 uniquely designed and textured Airsides (Terminals) connected with animated Monorails. Numerous static and animated objects like cars, trucks, vans, airport staff, etc.
Pub/Dev/Source:	Fly Tampa
Rec Spec:	Pentium III-1GHz or higher, 256 Mb or more RAM, 32 Mb 3D graphics card.
Platform:	FS2002 *FS2004
Type:	Download
Web Site:	http://www.flytampa.com

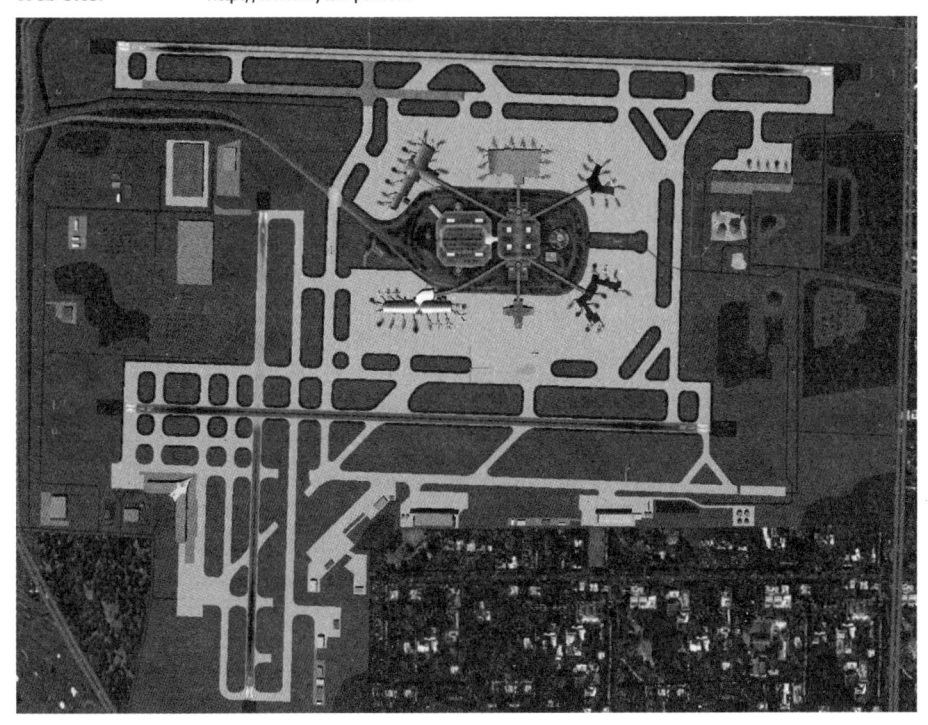

FS Add-On - Scenery   a-Z

Title: Fly to Hong Kong
TecPilot Rating: ✦✦✦✦✦
Description: The entire region of Hong Kong. Includes roadways, street signs, cars parked in garages, cars and buses clearly visible on streets throughout. The Ultimate Hong Kong Simulator brings you the entire Hong Kong region with over 67 Hong Kong Sightseeing Flights to help you discover every corner of this fabulous and exotic land.
Pub/Dev/Source: Aerosoft
Rec Spec: Pentium III-1GHz or higher, 256 Mb or more RAM, 32 Mb 3D graphics card.
Platform: FS2002 *FS2004
Type: Box
Web Site: http://www.aerosoft.com

Title: France Land Class
TecPilot Rating: ✦✦✦✦✦
Description: Covers the entire area of France. Land Class works with default or third party textures. Designed from real maps. The scenery draws all geographical features such as forests, fields, AutoGen scenery and cities. Exactly as they exist in in real life. Fully Compatible with other Terrain Mesh Sceneries.
Pub/Dev/Source: FSFreeware/Raimondo Taburet
Rec Spec: Pentium III-1GHz or higher, 256 Mb or more RAM, 32 Mb 3D graphics card.
Platform: FS2002 *FS2004
Type: Download
Web Site: http://www.fsfreeware.com

Title: France, Andorra, Channel Islands, Benelux
TecPilot Rating: ✦✦✦✦✦
Description: Depicting France and other terrain at high levels of 'mesh' detail - at low and high altitudes. This is "Gaia" (real earth mesh scenery). Can be merged with developers other products: Italy, Switzerland, Austria, Germany. Can be used on its own as a mesh scenery.
Pub/Dev/Source: FSFreeware/Raimondo Taburet
Rec Spec: Pentium III-1GHz or higher, 256 Mb or more RAM, 32 Mb 3D graphics card.
Platform: FS2000 FS2002 *FS2004
Type: Box or Download
Web Site: http://www.fsfreeware.com

Title: Frankfurt Intl (EDDF)
TecPilot Rating: ✦✦✦✦✦
Description: Includes Accurate and detailed scenery of Frankfurt Rhein-Main Intl Airport (EDDF) including buildings, runways, taxiways and airport objects. Realistic buildings and objects with full texturing. Active docking system and jet ways at each gate. serviceArmada(TM).
Pub/Dev/Source: SimFlyers Associated
Rec Spec: Pentium III-1GHz or higher, 256 Mb or more RAM, 32 Mb 3D graphics card.
Platform: FS2000 FS2002 *FS2004
Type: Download
Web Site: http://www.simflyers.net

a-z   FS Add-On - Scenery

Title:	**Frankfurt**
TecPilot Rating:	✦✦✦✦✦
Description:	Detailed design and many new great looking day and night time lighting effects. Summer, winter and spring textures. Fully 32 Bit compatible. Frame rate optimized. Dynamic service and apron cars. Static aircraft with night time lighting effects and automatic changing liv-
Pub/Dev/Source:	German Airports /Scenery Design Team
Rec Spec:	Pentium III-1GHz or higher, 256 Mb or more RAM, 32 Mb 3D graphics card.
Platform:	FS2000 FS2002 *FS2004
Type:	Download
Web Site:	http://www.germanairports.net

Title:	**Friederichshafen, Innsbruck Altenreign**
TecPilot Rating:	✦✦✦✦✦
Description:	Detailed design and many new great looking day and night time lighting effects. Summer, winter and spring textures. Fully 32 Bit compatible. Frame rate optimized. Dynamic service
Pub/Dev/Source:	German Airports /Scenery Design Team
Rec Spec:	Pentium III-1GHz or higher, 256 Mb or more RAM, 32 Mb 3D graphics card.
Platform:	FS2000 FS2002 *FS2004
Type:	Download
Web Site:	http://www.germanairports.net

Title:	**Fuerteventura Lanzarote 2003**
TecPilot Rating:	✦✦✦✦✦
Description:	Features high Resolution mesh terrain, photorealistic textures, night lightning, AI-traffic compatible, dynamic scenery, international airport "Fuerteventura (GCFV) and Lanzarote (GCRR).
Pub/Dev/Source:	SimWings/Thorsten Loth
Rec Spec:	Pentium III-1GHz or higher, 256 Mb or more RAM, 32 Mb 3D graphics card.
Platform:	FS2002 *FS2004
Type:	Download
Web Site:	http://www.sim-wings.de

Title:	**GB Airports 2001**
TecPilot Rating:	✦✦✦✦✦
Description:	GB Airports, is a continuation of English Airports for FS2000. GB include six of Great Britain's busiest airports, including; Glasgow, Edinburgh, Aberdeen, Newcastle, East Midlands and London City, all of which include true-to-life buildings, taxiways and even wind socks.
Pub/Dev/Source:	Barry Perfect
Rec Spec:	Pentium III-1GHz or higher, 256 Mb or more RAM, 32 Mb 3D graphics card.
Platform:	FS2000
Type:	Download
Web Site:	http://www.gbairports.co.uk

FS Add-On - Scenery   a-Z

Title: **GB Airports**
TecPilot Rating: ✦✦✦✦✦
Description: Each airport features accurate runways, helipads and taxiways - all with hold lines, CAT and ID signage and night lighting. Off the apron; buildings, accurately placed radar tow-

Pub/Dev/Source: Just Flight/Barry Perfect
Rec Spec: Pentium III-1GHz or higher, 256 Mb or more RAM, 32 Mb 3D graphics card.
Platform: FS2000 FS2002 *FS2004
Type: Box
Web Site: http://www.justflight.com

Title: **GeoRender 1 - Flying M Ranch & Ranger Creek**
TecPilot Rating: ✦✦✦✦✦
Description: Created by the well known scenery artist Richard Goldstein this is a complete remake of the highly acclaimed scenery of Flying M Ranch (OR) and Ranger Creek (WA) created for FS2000. The emphasis on this release is the accurate rendition of seasonal textures.

Pub/Dev/Source: Lago
Rec Spec: Pentium III-1GHz or higher, 256 Mb or more RAM, 32 Mb 3D graphics card.
Platform: FS2002 *FS2004
Type: Download
Web Site: http://www.lagoonline.com

Title: **GeoRender 2 - Bryce Canyon Airport**
TecPilot Rating: ✦✦✦✦✦
Description: Covers the high altitude Bryce Canyon Airport and its surrounding area. Serves Bryce Canyon and Garfield County and is owned by Garfield County. An ideal location to practice high altitude/high density flight procedures. Extended manual about procedures. Includes set of sounds made by MeatWater. Includes dynamic tracks and High Density Mesh of the surrounding area.

Pub/Dev/Source: Lago
Rec Spec: Pentium III-1GHz or higher, 256 Mb or more RAM, 32 Mb 3D graphics card.
Platform: FS2002 *FS2004
Type: Download
Web Site: http://www.lagoonline.com

Title: **GeoRender 3 - Diamond Point**
TecPilot Rating: ✦✦✦✦✦
Description: Covers Diamond Point Airstrip (2WA1) based in the beautiful state of Washington, eight miles North East of Sequim. This is the first GeoRender product made in GMax resulting in an increase in frame rates. The scenery has standard Active Scenery tracks and en-

Pub/Dev/Source: Lago
Rec Spec: Pentium III-1GHz or higher, 256 Mb or more RAM, 32 Mb 3D graphics card.
Platform: FS2002 *FS2004
Type: Download
Web Site: http://www.lagoonline.com

a-Z FS Add-On - Scen-

Title:	**German Airports 1**
TecPilot Rating:	✦✦✦✦✧
Description:	Contains eight German airports in high detail including Innsbruck and Altenrhein (CH) as a bonus. Munich, Franz-Josef Strauß Airport, Nürnberg, Dresden, Bayreuth, Egelsbach, Augsburg, Stuttgart, Friedrichshafen (Lake Bodensee). Innsbruck (A) and Altenrhein (CH). 'German Airports Action Scenery'. Adds a lot of first-class static and dynamic objects (moving docking bridges, follow-me-cars, busses, etc.).
Pub/Dev/Source:	Aerosoft
Rec Spec:	Pentium III-1GHz or higher, 256 Mb or more RAM, 32 Mb 3D graphics card.
Platform:	FS2000 FS2002 *FS2004
Type:	Box
Web Site:	http://www.aerosoft.com

SCENERY

Title:	**German Airports 2**
TecPilot Rating:	✦✦✦✦✧
Description:	Consists of 8 detailed airports. Frankfurt/Main, Leipzig/Halle, Kassel, Cologne/Bonn, Dortmund, Paderborn/Lippstadt, Muenster/Osnabrueck and Hannover. First class static and dynamic objects make the airports come alive. The smallest detail of each airport is recreated in three dimensions. Even equipment inside the cargo-bays can be seen.
Pub/Dev/Source:	Aerosoft
Rec Spec:	Pentium III-1GHz or higher, 256 Mb or more RAM, 32 Mb 3D graphics card.
Platform:	FS2000 FS2002 *FS2004
Type:	Box
Web Site:	http://www.aerosoft.com

Title:	**German Airports 3**
TecPilot Rating:	✦✦✦✦✧
Description:	Consists of 8 highly detailed airports. Duesseldorf, Mönchengladbach (Duesseldorf-Express-Airport) , Bremen, Hamburg, Kiel, Lübeck, , Erfurt Berlin-Tegel. First class night effects, static and dynamic objects and FSTraffic tracks. The smallest detail of each airport is displayed in three dimensions.
Pub/Dev/Source:	Aerosoft
Rec Spec:	Pentium III-1GHz or higher, 256 Mb or more RAM, 32 Mb 3D graphics card.
Platform:	FS2000 FS2002 *FS2004
Type:	Box
Web Site:	http://www.aerosoft.com

Join our online club today! We'll be happy to help!

WWW.TECPILOT.COM
Home Of "The Good Flight Simmer's Guide"

FS Add-On - Scenery   a-Z

Title: **German Airports 4**
TecPilot Rating: ✦✦✦✦✦
Description: Consists of 10 highly detailed airports including: Berlin-Tempelhof, Berlin-Schoenefeld, Sylt, Westerland, Magdeburg, Baden-Baden, Saarbruecken, Braunschweig Maastricht-Aachen and Hahn. The smallest detail of each airport is displayed. Even the equipment inside the cargo-bays can be seen.

Pub/Dev/Source: Aerosoft
Rec Spec: Pentium III-1GHz or higher, 256 Mb or more RAM, 32 Mb 3D graphics card.
Platform: FS2000 FS2002 *FS2004
Type: Box
Web Site: http://www.aerosoft.com

Title: **German Airports Selection 2002**
TecPilot Rating: ✦✦✦✦✦
Description: Each airport is displayed in three-dimension with characteristic buildings. Many detailed static and dynamic objects like moving docking bridges, follow-me-cars, busses and refuelling trucks create a lot of atmosphere. Even equipment inside the hangars can be seen. Spectators and flight staff look outside the buildings to watch the air traffic. This scenery is fully compatible with the new AI-Traffic of FS2002. It is also compatible with Flight Simulator 2000.

Pub/Dev/Source: Aerosoft
Rec Spec: Pentium III-1GHz or higher, 256 Mb or more RAM, 32 Mb 3D graphics card.
Platform: FS2000 FS2002 *FS2004
Type: Box
Web Site: http://www.aerosoft.com

Title: **Germany Land Class**
TecPilot Rating: ✦✦✦✦✦
Description: Covers the whole area of Germany. Land Class works with default or third party textures. Designed from real maps, the scenery draws all geographical features such as forests, fields, AutoGen scenery and cities. Exactly as they exist in in real life. Fully Compatible with other Terrain Mesh Sceneries.

Pub/Dev/Source: FSFreeware/Raimondo Taburet
Rec Spec: Pentium III-1GHz or higher, 256 Mb or more RAM, 32 Mb 3D graphics card.
Platform: FS2002 *FS2004
Type: Download
Web Site: http://www.fsfreeware.com

**2500 members from over 60 Countries use TecPilot weekly.
Over 1 million people visit TecPilot for their sim news every year!**

WWW.TECPILOT.COM

Home Of "The Good Flight Simmer's Guide"

a-Z   FS Add-On - Scenery

Title: **Global Topographic Enhancement**
TecPilot Rating: ✦✦✦✦✦
Description: GTE enhances "ome" areas with a 10 meters resolution. The Alps, Azores, Balearic Islands, UK, California, Northern Europe, Hawaii, Cape Horn, South-west Europe, France, and some more. The effect is not very significant. (According to the Aerosoft web site) but we think its an excellent addition to the FS arsenal - especially the UK where it really needs it!
Pub/Dev/Source: Aerosoft
Rec Spec: Pentium III-1GHz or higher, 256 Mb or more RAM, 32 Mb 3D graphics card.
Platform: FS2002 *FS2004
Type: Box
Web Site: http://www.aerosoft.com

Title: **Goiania and Anapolis (Brazil Singles)**
TecPilot Rating: ✦✦✦✦✦
Description: Scenery features: SBGO - Santa Genoveva Airport", in Goiania. SBXN - Aeroporto Municipal de Anápolis", in Anápolis. SWNV - Aeródromo Nacional da Aviação", in Goiania. Realistic buildings and objects with full texturing. AI traffic compatible. Documentation with airport charts (taxi, departure, approach). Traffic files to be used with TTools reproducing real flights in the airport.
Pub/Dev/Source: RealFlight
Rec Spec: Pentium III-1GHz or higher, 256 Mb or more RAM, 32 Mb 3D graphics card.
Platform: FS2002 *FS2004
Type: Download
Web Site: http://www.realflight.com.br

Title: **Gothenburg Säve 2002**
TecPilot Rating: ✦✦✦✦✦
Description: Säve airport is the second largest airport in Gothenburg. Ryanair is seen arriving and departing from here regularly. This scenery package features photo realistic (effect) buildings, static and dynamic scenery and seasonal effects. Car parking is rendered behind the
Pub/Dev/Source: FS Dream Factory/Cornel Grigoriu
Rec Spec: Pentium III-1GHz or higher, 256 Mb or more RAM, 32 Mb 3D graphics card.
Platform: FS2002 *FS2004
Type: Download
Web Site: http://www.fsdreamfactory.com

Title: **Gran Canaria 2003**
TecPilot Rating: ✦✦✦✦✦
Description: Features high resolution mesh terrain, photorealistic textures, night lightning, AI-traffic compatible, dynamic scenery, international airport "GCLP" and the aeroclub "El Berriel
Pub/Dev/Source: SimWings/Thorsten Loth
Rec Spec: Pentium III-1GHz or higher, 256 Mb or more RAM, 32 Mb 3D graphics card.
Platform: FS2002 *FS2004
Type: Download
Web Site: http://www.sim-wings.de

FS Add-On - Scenery   a-Z

Title:	**Grand Canyon 10m Terrain**
TecPilot Rating:	✦✦✦✦✦
Description:	The Grand Canyon in its entirety from Lake Powell to Lake Mead in 9.6m (LOD=12). Re-sampled directly from USGS 10m digital elevation models. Includes a significantly larger surrounding area of 19.2m (LOD=11) coverage. Impressive!
Pub/Dev/Source:	FSGenesis
Rec Spec:	Pentium III-1GHz or higher, 256 Mb or more RAM, 32 Mb 3D graphics card.
Platform:	FS2002 *FS2004
Type:	Box
Web Site:	http://www.fsgenesis.com

Title:	**Grand Canyon, Hawaii, Yellowstone, Guam**
TecPilot Rating:	✦✦✦✦✦
Description:	First pack in a series. Includes Four separate mesh sceneries: Hawaii, Yellowstone National Park, The Grand Canyon of Arizona, Guam Island in the Pacific. Designed with USGS 10 meter (DEM) resolution. We recommend this for VFR aviators only.
Pub/Dev/Source:	FSFreeware/Raimondo Taburet
Rec Spec:	Pentium III-1GHz or higher, 256 Mb or more RAM, 32 Mb 3D graphics card.
Platform:	FS2002 *FS2004
Type:	Box or Download
Web Site:	http://www.fsfreeware.com

Title:	**Great Britain & Ireland**
TecPilot Rating:	✦✦✦✦✦
Description:	Great Britain and Ireland contains over 280 highly detailed renditions of airports located around the British Isles. From international airports like Heathrow, Shannon and Manchester to smaller airports like Biggin Hill and Duxford, right down to local strips such as Kilkenny and Halfpenny Green. Everything in this collection has been carefully researched over 18 months in order to reproduce it at the highest level of detail.
Pub/Dev/Source:	Just Flight / Barry Perfect
Rec Spec:	Pentium III 700 or higher, 128Mb or more RAM, 32Mb 3D graphics accelerator card, Monitor capable of 1024 x 768 resolution, 3D sound card
Platform:	FS2000
Type:	Box
Web Site:	http://www.aerosoft.com

For the very latest downloads including
scenery, aircraft and utilities visit...

WWW.SIMMARKET.COM
Your one stop flight shop!

a-Z

FS Add-On - Scenery

Title:	**Greece & Balcans Land Class**
TecPilot Rating:	✦ ✦ ✦ ✦ ✦
Description:	Covers the whole area of Greece and the Balcans. Land Class works with default or third party textures. Designed from real maps, the scenery draws all geographical features such as forests, fields, AutoGen scenery and cities. Exactly as they exist in in real life. Fully Compatible with other Terrain Mesh Sceneries.
Pub/Dev/Source:	FSFreeware/Raimondo Taburet
Rec Spec:	Pentium III-1GHz or higher, 256 Mb or more RAM, 32 Mb 3D graphics card.
Platform:	FS2002 *FS2004
Type:	Download
Web Site:	http://www.fsfreeware.com

Title:	**Greece, Turkey and Middle-East**
TecPilot Rating:	✦ ✦ ✦ ✦ ✦
Description:	This mesh scenery boasts complete coverage of: Greece, Turkey, Northern Iraq, Syria, NW Georgia and the Lebanon. Designed at 1 Arc/Sec resolution.
Pub/Dev/Source:	FSFreeware/Raimondo Taburet
Rec Spec:	Pentium III-1GHz or higher, 256 Mb or more RAM, 32 Mb 3D graphics card.
Platform:	FS2000 FS2002 *FS2004
Type:	Box or Download
Web Site:	http://www.fsfreeware.com

Title:	**Hamburg (EDDH) + Bremen (EDDW)**
TecPilot Rating:	✦ ✦ ✦ ✦ ✦
Description:	Detailed design and many new great looking day and night time lighting effects. Summer, winter and spring textures. Fully 32 Bit compatible. Frame rate optimized. Dynamic service and apron cars. Static aircraft with night time lighting effects and automatic changing liveries. Runways with wheel skid marks. New Terminals, runways, towers and other buildings. SAFEGATE docking system and active dock bridges.
Pub/Dev/Source:	German Airports /Scenery Design Team
Rec Spec:	Pentium III-1GHz or higher, 256 Mb or more RAM, 32 Mb 3D graphics card.
Platform:	FS2000 FS2002 *FS2004
Type:	Download
Web Site:	http://www.germanairports.net

Title:	**Hawaiian Islands 9.6m Terrain & Landclass**
TecPilot Rating:	✦ ✦ ✦ ✦ ✦
Description:	Hawaiian Islands 10m Terrain & Landclass covering the islands of Hawaii, Kahoolawe, Maiu, Lanai, Molokai, Oahu, Kauai, and Niihau. Terrain rendered from USGS 10m source files. Landclass derived from NOAA C-CAP dataset (2000). 65 Mb download. .
Pub/Dev/Source:	FSGenesis
Rec Spec:	Pentium III-1GHz or higher, 256 Mb or more RAM, 32 Mb 3D graphics card.
Platform:	FS2002 *FS2004
Type:	Download
Web Site:	http://www.fsgenesis.com

FS Add-On - Scenery a-Z

Title: **Himalaja**
TecPilot Rating: ✈✈✈✈✈
Description: Includes the whole area of Nepal and airports with special structures and Katmandu International airport near to the highest mountains in the world. Additionally, Tibet China, parts of India, Pakistan and Bhutan ! 34 airports in total in the mountains - some of them are very dangerous ! See Mount Everest or K2 or look for the Yeti. You will be surprised!
Pub/Dev/Source: AriFlight
Rec Spec: Pentium III-1GHz or higher, 256 Mb or more RAM, 32 Mb 3D graphics card.
Platform: FS98
Type: Download
Web Site: http://www.ari.de/

Title: **Hong Kong**
TecPilot Rating: ✈✈✈✈✈
Description: All taxiway and RWY lights included. All Taxiway signs are night illuminated. 2 Docking Gates and static aircraft. All buildings constructed or opened in Summer 1999. ILS and NAV's for RWY 07R/25L (RWY07L/25R was/is under construction). Includes an Airport Chart showing the Airport under construction. Whoopee!
Pub/Dev/Source: Skysoft / Holger Schmidt
Rec Spec: Pentium 500, 64 MByte RAM, 3D Graphics Card.
Platform: FS95 FS98 FS2000
Type: Download
Web Site: http://www.simmarket.com/online

Title: **Honolulu**
TecPilot Rating: ✈✈✈✈✈
Description: Written using Gmax. The scenery is fast! Comes with its own AI traffic and 4 aircraft (not flyable). These are included to fly scheduled flights to and from this airport. Also incorporated are a set of high resolution charts that feature the curved approach to runway 26 plus VOR/GPS and NDB/GPS data.
Pub/Dev/Source: Lago
Rec Spec: Pentium III-1GHz or higher, 256 Mb or more RAM, 32 Mb 3D graphics card.
Platform: FS2002 *FS2004
Type: Download
Web Site: http://www.lagoonline.com

Title: **Ibiza 2001 v2**
TecPilot Rating: ✈✈✈✈✈
Description: The scenery covers the airport of Ibizia. It replaces default scenery with photorealistic buildings and night lighting effects. Not an amazing add-on but we have heard good reports about it from flight simmer's online.
Pub/Dev/Source: Sim-Wings FS Software
Rec Spec: Pentium 500, 64 MByte RAM, 3D Graphics Card.
Platform: FS2000
Type: Download
Web Site: http://www.simmarket.com/online

a-Z

FS Add-On - Scenery

Title:	**Islands of the West Indies: Jewels of the Caribbean v1**
TecPilot Rating:	✦ ✦ ✦ ✦ ✦
Description:	The first release of the Jewels of the Caribbean Series, Islands of the West Indies Volume 1, features "mesh terrain" elevation data, hand coloured and rendered bitmaps, extensive night lighting effects and custom renditions of all the airports in the region. The islands of Anguilla, St.Maarten/St.Martin, St.Barts, Saba, St.Eustatius, St.Kitts and Nevis, Montserrat,
Pub/Dev/Source:	Flight 1
Rec Spec:	Pentium III-1GHz or higher, 256 Mb or more RAM, 32 Mb 3D graphics card.
Platform:	FS2002 *FS2004
Type:	Download
Web Site:	http://www.flight1.com

Title:	**Islands of the West Indies: Jewels of the Caribbean v2**
TecPilot Rating:	✦ ✦ ✦ ✦ ✦
Description:	The second release of the Jewels of the Caribbean Series, Islands of the West Indies Volume 2, features "mesh terrain" elevation data, hand coloured and rendered bitmaps, night lighting effects and custom renditions of all the airports in the region. The islands of Guadeloupe, La Desirade, Marie Galante, Iles Saintes, Dominica, Martinique and St.Lucia are covered.
Pub/Dev/Source:	Flight 1
Rec Spec:	Pentium III-1GHz or higher, 256 Mb or more RAM, 32 Mb 3D graphics card.
Platform:	FS2002 *FS2004
Type:	Download
Web Site:	http://www.flight1.com

Title:	**Italy 2000**
TecPilot Rating:	✦ ✦ ✦ ✦ ✦
Description:	The included scenery uses the FS2000 default territory as its base, enhancing the existing airports by replacing the default buildings, adding new ones as well as the areas adjacent the airports. Also includes a set of tracks (moving aircraft) that fly the main proce-
Pub/Dev/Source:	Lago
Rec Spec:	Pentium III-1GHz or higher, 256 Mb or more RAM, 32 Mb 3D graphics card.
Platform:	FS2000 FS2002 *FS2004
Type:	Box
Web Site:	http//www.lagoonline.com

What do you call a person who knows where
to get up-to-date and relevant information
about Flight Simulator Add-On Products?

Right, a WWW.TECPILOT.COM member!

FS Add-On - Scenery   a-Z

Title: **Italy, Tunisia, Malta & Corse**
TecPilot Rating: ✦✦✦✦✦
Description: A mesh scenery reproduced at 40-50 meter resolution. Covers Italy, Tunisia, Malta & Corse. Can be merged with developers other products: Switzerland, Austria and Germany, for example.
Pub/Dev/Source: FSFreeware/Raimondo Taburet
Rec Spec: Pentium III-1GHz or higher, 256 Mb or more RAM, 32 Mb 3D graphics card.
Platform: FS2000 FS2002 *FS2004
Type: Box or Download
Web Site: http://www.fsfreeware.com

Title: **Japan Land Class**
TecPilot Rating: ✦✦✦✦✦
Description: Covers the whole area of Japan. Land Class works with default or third party textures. Designed from real maps the scenery draws all geographical features such as forests, fields, AutoGen scenery and cities. Exactly as they exist in in real life. Fully compatible with other Terrain Mesh Sceneries.
Pub/Dev/Source: FSFreeware/Raimondo Taburet
Rec Spec: Pentium III-1GHz or higher, 256 Mb or more RAM, 32 Mb 3D graphics card.
Platform: FS2002 *FS2004
Type: Download
Web Site: http://www.fsfreeware.com

Title: **Jonkoping Airport**
TecPilot Rating: ✦✦✦✦✦
Description: Featuring a highly detailed and accurate rendition of one of the regional airports in western Sweden. The airport serves with flights to Swedish destinations as well as some charter traffic to southern Europe. Buildings with photo realistic textures. Mesh scenery around the airport. Static scenery. AI compatible. Fully functional Wig-wags.
Pub/Dev/Source: FS Dream Factory/Cornel Grigoriu
Rec Spec: Pentium III-1GHz or higher, 256 Mb or more RAM, 32 Mb 3D graphics card.
Platform: FS2002 *FS2004
Type: Download
Web Site: http://www.fsdreamfactory.com

Title: **Kiruna 2002**
TecPilot Rating: ✦✦✦✦✦
Description: A scenery covering Kiruna Airport (ESNQ) and part of the city of Kiruna which includes all major landmarks. Includes Photo realistic buildings. Every building is represented at the airport using high resolution textures. Static and dynamic scenery. Seasonal effects on buildings and ground with snow in the winter as it is in this arctic region. Night effects.
Pub/Dev/Source: FS Dream Factory/Cornel Grigoriu
Rec Spec: Pentium III-1GHz or higher, 256 Mb or more RAM, 32 Mb 3D graphics card.
Platform: FS2000 FS2002 *FS2004
Type: Download
Web Site: http://www.fsdreamfactory.com

a-Z   FS Add-On - Scenery

Title:	**Konotop AFB**
TecPilot Rating:	✦✦✦✦✦
Description:	The Konotop Air Force Base is located near the state border between Russia Ukraine and Belorus. The base is one of eight training facilities of Chernigov Higher Fighter Pilot School. 105-th Fighter Pilots Training Wing and Military Helicopter Repair Plant are dislocated at the base.
Pub/Dev/Source:	CaptainSim
Rec Spec:	Pentium III-1GHz or higher, 256 Mb or more RAM, 32 Mb 3D graphics card.
Platform:	FS2000 FS2002 CFS2
Type:	Download
Web Site:	http://www.aerol39.com

Title:	**Kuala Lumpur Sepang Airport**
TecPilot Rating:	✦✦✦✦✦
Description:	Revamp of original FS2000 Scenery. Includes the main terminal building with contact pier, satellite building and gangways at all the gates. Docking system, taxiways with centrelines and signs and all runways with Navaida are included. Parking facilities, freight buildings and hangars, night lighting apron, dynamic scenery with the people mover, 24 planes and 10 catering trucks complete the picture.
Pub/Dev/Source:	MICE Innovative Computer Engineering/Hans-Jörg Müller
Rec Spec:	Pentium III-1GHz or higher, 256 Mb or more RAM, 32 Mb 3D graphics card.
Platform:	FS2000 FS2002 *FS2004
Type:	Download
Web Site:	http://www.mice.ch

Title:	**Landvetter 2002**
TecPilot Rating:	✦✦✦✦✦
Description:	Part of the Scandinadian Series. Includes Photo realistic buildings using high resolution textures. Static and dynamic scenery. Seasonal effects. Car parking area behind the terminal made with hand painted polygons. Sign system on the taxiways, using real charts as reference. Advanced ground lighting. Static scenery. Dynamic scenery with baggage cars, busses and a catering trucks.
Pub/Dev/Source:	FS Dream Factory/Cornel Grigoriu
Rec Spec:	Pentium III-1GHz or higher, 256 Mb or more RAM, 32 Mb 3D graphics card.
Platform:	FS2000 FS2002 *FS2004
Type:	Download
Web Site:	http://www.fsdreamfactory.com

Title:	**Le Monde de Chapsendor**
TecPilot Rating:	✦✦✦✦✦
Description:	The imaginary world of "Le Monde de Chapsendor". Made of 4 islands, it is placed to the West of the island of Corsica. The first of this isles is "Manenchito", the most important and the place of departure of this scenery environment. Every island benefits from 3D enhancements, Static aircraft and dynamic devices.
Pub/Dev/Source:	Epilogic Aviation/Patrice Kuczynski & Michel Philizot
Rec Spec:	Pentium III-1GHz or higher, 256 Mb or more RAM, 32 Mb 3D graphics card.
Platform:	FS2000 (free) FS2002 *FS2004
Type:	Download
Web Site:	http://www.epilogic-aviation.net

FS Add-On - Scenery   a-Z

Title: **Leipzig (EDDP)**
Hannover (EDDV)
TecPilot Rating: ✈✈✈✈✈
Description: Detailed design and many new great looking day and night time lighting effects. Summer, winter and spring textures. Fully 32 Bit compatible. Frame rate optimized. Dynamic service and apron cars. Static aircraft with night time lighting effects and automatic changing liveries. Runways with wheel skid marks.

Pub/Dev/Source: German Airports /Scenery Design Team
Rec Spec: Pentium III-1GHz or higher, 256 Mb or more RAM, 32 Mb 3D graphics card.
Platform: FS2000 FS2002 *FS2004
Type: Download
Web Site: http://www.germanairports.net

Title: **London Airports**
TecPilot Rating: ✈✈✈✈✈
Description: London Airports is a collection of 5 airports in the London area which include Heathrow, Stansted, Gatwick, Luton and London City. This scenery represents one of the finest flight sim sceneries you will ever find. Just make sure you have a PC that can run it!

Pub/Dev/Source: UK2000 Scenery/Gary Summons
Rec Spec: Pentium III-1GHz or higher, 256 Mb or more RAM, 32 Mb 3D graphics card.
Platform: FS2000 FS2002 *FS2004
Type: Download
Web Site: http://www.uk2000scenery.com

Title: **London Gatwick (EGKK)**
TecPilot Rating: ✈✈✈✈✈
Description: Stunning scenery of this major UK Airport, Includes moving air gates, accurate buildings, every parking bay, taxiway centre lines, signs, breath taking night effects, static aircraft, dynamic objects and all this at a good frame rate!.

Pub/Dev/Source: UK2000 Scenery/Gary Summons
Rec Spec: Pentium III-1GHz or higher, 256 Mb or more RAM, 32 Mb 3D graphics card.
Platform: FS2000 FS2002 *FS2004
Type: Download
Web Site: http://www.uk2000scenery.com

Title: **London Heathrow (EGLL)**
TecPilot Rating: ✈✈✈✈✈
Description: Also included in the London Airports package (above). Features buildings which are accurately modelled and textured. Includes taxiway centrelines, all stands and air gates, timed departure boards, timed stop bars, passengers gathering prior to the departure time, taxiway/stand markings that are realistic and, and, we simply haven't got enough space to explain it all!.

Pub/Dev/Source: UK2000 Scenery/Gary Summons
Rec Spec: Pentium III-1GHz or higher, 256 Mb or more RAM, 32 Mb 3D graphics card.
Platform: FS2000 FS2002 *FS2004
Type: Download
Web Site: http://www.uk2000scenery.com

a-z   FS Add-On - Scenery

Title: **Los Angeles Intl (KLAX)**
TecPilot Rating: ✦ ✦ ✦ ✦ ✦
Description: Includes accurate and detailed scenery of Los Angeles Intl (KLAX) including buildings, runways, taxiways and airport objects. Realistic buildings and objects with full texturing. Active docking system and jet ways at each gate. serviceArmada(TM).
Pub/Dev/Source: SimFlyers Associated
Rec Spec: Pentium III-1GHz or higher, 256 Mb or more RAM, 32 Mb 3D graphics card.
Platform: FS2000 FS2002 *FS2004
Type: Download
Web Site: http://www.simflyers.net

Title: **Luxembourg Airports**
TecPilot Rating: ✦ ✦ ✦ ✦ ✦
Description: A complete new design that provides state-of-the-art airport and scenery. The airports, ELLX and ELNT not only come with a multitude of related objects but also incorporate: Dynamic airport traffic (busses, cars, trucks and kerosene transporters). Airport service vehicles and personnel on the apron. Parking wardens in both passenger and freight terminals, highly detailed, photo-textured aircraft with night-illumination and adapted AI-traffic.
Pub/Dev/Source: Aerosoft
Rec Spec: Pentium III-1GHz or higher, 256 Mb or more RAM, 32 Mb 3D graphics card.
Platform: FS2002 *FS2004
Type: Box
Web Site: http://www.aerosoft.com

Title: **Madeira 2002**
TecPilot Rating: ✦ ✦ ✦ ✦ ✦
Description: Based on Funchal on the Island of Madeira. Regarded as one of the most dangerous airport approaches in spite of a restoration project in 2001. Half of the runway is built on pylons straddling deep water and rocks.
Pub/Dev/Source: Airsim/Jacques Sylvain
Rec Spec: Pentium III-1GHz or higher, 256 Mb or more RAM, 32 Mb 3D graphics card.
Platform: FS2002 *FS2004
Type: Download
Web Site: http://www.airsim.free.fr

Title: **Malaga 2003**
TecPilot Rating: ✦ ✦ ✦ ✦ ✦
Description: The scenery contains the international airport of Malaga (LEMG) with airport museum, old terminal buildings and military section. A PDF manual in German & English is provided which covers installation and airport history. Static aircraft can be removed if you prefer standard AI-traffic.
Pub/Dev/Source: SimWings/Thorsten Loth
Rec Spec: Pentium III-1GHz or higher, 256 Mb or more RAM, 32 Mb 3D graphics card.
Platform: FS2002 *FS2004
Type: Download
Web Site: http://www.sim-wings.de

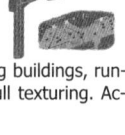

FS Add-On - Scenery   a-Z

Title: **Mallorca 2002**
TecPilot Rating: ✈✈✈✈✈
Description: Comes with two detailed airports: Son San Joan and Son Bonet. Includes the port of Palma, Playa de Palma (Arenal) and high resolution mesh. Features photorealistic tex-

Pub/Dev/Source: SimWings/Thorsten Loth
Rec Spec: Pentium III-1GHz or higher, 256 Mb or more RAM, 32 Mb 3D graphics card.
Platform: FS2002 *FS2004
Type: Download
Web Site: http://www.sim-wings.de

Title: **Malmoe (ESMS) & Gothenborg (ESGG)**
TecPilot Rating: ✈✈✈✈✈
Description: Fully seasonal and night ground texturing. Most buildings are exquisitely photo textured in 16 bit Hi-res and night lighted. Features CAA high scale drawings used for the airport layouts. Latest navigational information and databases used. Full real-life SAFEGATE®, APIS++® and APIS® Docking Systems at all relevant gates. Instruction file in PDF-

Pub/Dev/Source: SwedFlight Pro
Rec Spec: Pentium III-1GHz or higher, 256 Mb or more RAM, 32 Mb 3D graphics card.
Platform: FS2002 *FS2004
Type: Download
Web Site: http://www.swedflight.com

Title: **Manaus - Eduardo Gomes Intl (SBEG)**
TecPilot Rating: ✈✈✈✈✈
Description: Detailed objects such as detailed steps on the gate staircase included. Full night illumination. Uses the most of the FS2002 enhancements to raise FPS. Carefully textures hangars and buildings. AFCAD import file included. FS2002 style centrelines and edge

Pub/Dev/Source: Flight Sim Brazil/Leandro Machado, Carlos Pereira
Rec Spec: Pentium III-1GHz or higher, 256 Mb or more RAM, 32 Mb 3D graphics card.
Platform: FS2002 *FS2004
Type: Download
Web Site: http://www.flightsimbrasil.comflytobrazil2

Title: **Manchester (EGCC)**
TecPilot Rating: ✈✈✈✈✈
Description: Features a new style interactive air gate system with 6 (yes SIX) moving parts, night lighting, static ground vehicles and aircraft, dynamic ground vehicles and aircraft, food truck, working stop bars, lead off light, Guard lights. All textures made to the highest

Pub/Dev/Source: UK2000 Scenery/Gary Summons
Rec Spec: Pentium III-1GHz or higher, 256 Mb or more RAM, 32 Mb 3D graphics card.
Platform: FS2000 FS2002 *FS2004
Type: Download
Web Site: http://www.uk2000scenery.com

The Good Flight Simmer's Guide Mk.II - 2003

a-z   FS Add-On - Scenery

Title: **Manchester Intl (EGCC)**
TecPilot Rating: ✦✦✦✦✦
Description: Accurate and detailed scenery of Manchester Intl (EGCC) which incluldes buildings, runways, taxiways and airport objects. Full texturing. Active docking system and jet ways at each gate. Note, this scenery is **pre**-second runway. serviceArmada(TM).
Pub/Dev/Source: SimFlyers Associated
Rec Spec: Pentium III-1GHz or higher, 256 Mb or more RAM, 32 Mb 3D graphics card.
Platform: FS2000 FS2002 *FS2004
Type: Download
Web Site: http://www.simflyers.net

Title: **Menorca 2001/2**
TecPilot Rating: ✦✦✦✦✦
Description: Comes with a detailed international airport that includes static aircraft and dynamic objects. Also contains the "Aero club de Mahon", the city of Ciutadella and hotels. The coastline was edited to match the real thing and is guarded by lighthouses. Includes photorealistic textures and night lightning.
Pub/Dev/Source: SimWings/Manfred Spatz
Rec Spec: Pentium III-1GHz or higher, 256 Mb or more RAM, 32 Mb 3D graphics card.
Platform: FS2002 *FS2004
Type: Download
Web Site: http://www.sim-wings.de

Title: **Milan Airports 2002**
TecPilot Rating: ✦✦✦✦✦
Description: This product includes an update of the three Milan airports that feature in Angelo Moneta's "Italy 2000" scenery. These airports vary form big the Hubs: Malpensa and Linate to the small Bresso airport, home of the local aeroclub.
Pub/Dev/Source: Lago
Rec Spec: Pentium III-1GHz or higher, 256 Mb or more RAM, 32 Mb 3D graphics card.
Platform: FS2002 *FS2004
Type: Download
Web Site: http://www.lagoonline.com

Title: **Monument Valley 10m Terrain**
TecPilot Rating: ✦✦✦✦✦
Description: Resampled directly from USGS 10m digital elevation models to 9.6m (LOD=12), this terrain CD is the first of it's kind to cover such a large area with such highly-detailed terrain. Depicting one of the more spectacular geographical areas of the United States, Monument Valley straddles northern Arizona and southern Utah, encompassing nearly one square degree in 9.6m resolution, surrounded by a significantly larger area covered in 19.1m resolution.
Pub/Dev/Source: FSGenesis
Rec Spec: Pentium III-1GHz or higher, 256 Mb or more RAM, 32 Mb 3D graphics card.
Platform: FS2002 *FS2004
Type: Box
Web Site: http://www.fsgenesis.com

FS Add-On - Scenery   a-Z

Title:	**Munich (EDDM)**
TecPilot Rating:	✦✦✦✦✦
Description:	Detailed design and many new great looking day and night time lighting effects. Summer, winter and spring textures. Fully 32 Bit compatible. Frame rate optimized. Dynamic service and apron cars. Static aircraft with night time lighting effects and automatic changing liveries. Runways with wheel skid marks.
Pub/Dev/Source:	German Airports /Scenery Design Team
Rec Spec:	Pentium III-1GHz or higher, 256 Mb or more RAM, 32 Mb 3D graphics card.
Platform:	FS2000 FS2002 *FS2004
Type:	Download
Web Site:	http://www.germanairports.net

Title:	**National Parks**
TecPilot Rating:	✦✦✦✦✦
Description:	National Parks Enhanced Scenery transforms 53 majestic places into even more vivid, visual memories. From the rich green hillsides of the Appalachians, the rugged slopes of Rocky Mountain or the deep and winding canyons of the Colorado, you'll soon enjoy them in super high detail. Three types of the DH2 Beaver (wheels, skis and float) let you fly from a variety of surfaces.
Pub/Dev/Source:	Abacus
Rec Spec:	Pentium III-1GHz or higher, 256 Mb or more RAM, 32 Mb 3D graphics card.
Platform:	FS2000 FS2002
Type:	Box
Web Site:	http://www.abacuspub.com

Title:	**New Zealand**
TecPilot Rating:	✦✦✦✦✦
Description:	Covers New Zealand in mesh reproduced at 40-50 meter resolution. A demo is available at 181Kb. The product file size is minimal and a tad expensive. We recommend purchasing the download version and not the CD.
Pub/Dev/Source:	FSFreeware/Raimondo Taburet
Rec Spec:	Pentium III-1GHz or higher, 256 Mb or more RAM, 32 Mb 3D graphics card.
Platform:	FS2000 FS2002 *FS2004
Type:	Box or Download
Web Site:	http://www.fsfreeware.com

Title:	**Newark Intl (KEWR)**
TecPilot Rating:	✦✦✦✦
Description:	Includes Accurate and detailed scenery of Newark Intl KEWR including buildings, runways, taxiways and airport objects. Realistic buildings and objects with full texturing. Active docking system and jet ways at each gate. serviceArmada(TM).
Pub/Dev/Source:	SimFlyers Associated
Rec Spec:	Pentium III-1GHz or higher, 256 Mb or more RAM, 32 Mb 3D graphics card.
Platform:	FS2000 FS2002 *FS2004
Type:	Download
Web Site:	http://www.simflyers.net

a-Z   FS Add-On - Scenery

Title:	**Nurnberg, Stuttgart**
TecPilot Rating:	✦✦✦✦✦
Description:	Detailed design and many new great looking day and night time lighting effects. Summer, winter and spring textures. Fully 32 Bit compatible. Frame rate optimized. Dynamic service and apron cars. Static aircraft with night time lighting effects and automatic changing liveries. Runways with wheel skid marks.
Pub/Dev/Source:	German Airports /Scenery Design Team
Rec Spec:	Pentium III-1GHz or higher, 256 Mb or more RAM, 32 Mb 3D graphics card.
Platform:	FS2000 FS2002 *FS2004
Type:	Download
Web Site:	http://www.germanairports.net

Title:	**Orlando Intl (KMCO)**
TecPilot Rating:	✦✦✦✦✦
Description:	Includes Accurate and detailed scenery of Orlando Intl KMCO including buildings, runways, taxiways and airport objects. Realistic buildings and objects with full texturing. Active docking system and jetways at each gate. serviceArmada(TM) at each terminal gate.
Pub/Dev/Source:	SimFlyers Associated
Rec Spec:	Pentium III-1GHz or higher, 256 Mb or more RAM, 32 Mb 3D graphics card.
Platform:	FS2000 FS2002 *FS2004
Type:	Download
Web Site:	http://www.simflyers.net

Title:	**Oslo Gardermoen Airport**
TecPilot Rating:	✦✦✦✦✦
Description:	The sixth airport in the series of Scandinavian Airports which covers Norway. Features a highly detailed (and accurate) rendition of one of the bigger hubs in Scandinavia. Oslo Gardermoen is situated in the south east part of the beautiful country of Norway. .
Pub/Dev/Source:	FS Dream Factory/Cornel Grigoriu
Rec Spec:	Pentium III-1GHz or higher, 256 Mb or more RAM, 32 Mb 3D graphics card.
Platform:	FS2000
Type:	Download
Web Site:	http://www.fsdreamfactory.com

Title:	**Pacific Northwest**
TecPilot Rating:	✦✦✦✦✦
Description:	The Northwest United States gets a facelift in PC Aviator's Pacific Northwest. The scenery 65,000 miles with hi-resolution photo-realistic mesh terrain. Seattle, Portland and Vancouver are also included, and drawn in 5 meter per pixel resolution.
Pub/Dev/Source:	PC Aviator
Rec Spec:	Pentium 200 (recommended), 32 Mb Ram, Microsoft Flight Simulator 98, CD Rom Drive, SoundBlaster Compatible Sound Card, 2D/3D Card, Headphones or Speakers recommended.
Platform:	FS98 CFS
Type:	Box
Web Site:	http://www.pcaviator.com

The Good Flight Simmer's Guide Mk.II - 2003   199

FS Add-On - Scenery   a-Z

Title: **Palma de Mallorca 'Son Sant Joan' Airport**
TecPilot Rating: ✈✈✈✈✈
Description: Son Sant Joan, the biggest charter airport in Europe, on the island of Mallorca in the Mediterranean Sea is available for FS2000 and FS2002 in German Airports/STD quality with fantastic night textures and all the Navaids you need.
Pub/Dev/Source: SDT - Scenery Design Team/Michael Kopatz
Rec Spec: Pentium III-1GHz or higher, 256 Mb or more RAM, 32 Mb 3D graphics card.
Platform: FS2000 FS2002 *FS2004
Type: Download
Web Site: http://www.dus2001.de

Title: **Pamplona 2001/2**
TecPilot Rating: ✈✈✈✈✈
Description: This scenery offers Pamplona's airport, the town of Pamplona itself (with the arena), the Ciudadela and the "Plaza Castillio".
Pub/Dev/Source: SimWings/Manfred Spatz
Rec Spec: Pentium III-1GHz or higher, 256 Mb or more RAM, 32 Mb 3D graphics card.
Platform: FS2002 *FS2004
Type: Download
Web Site: http://www.sim-wings.de

Title: **Paraíba and Santa Catarina**
TecPilot Rating: ✈✈✈✈✈
Description: Features carefully detailed and placed 3d objects, night illumination, AI traffic layout (Import via AFCAD). Uses FS2002's advanced resources. Charts included in PDF format (Adobe Acrobat).
Pub/Dev/Source: Flight Sim Brazil/Leandro Machado, Carlos Pereira
Rec Spec: Pentium III-1GHz or higher, 256 Mb or more RAM, 32 Mb 3D graphics card.
Platform: FS2002 *FS2004
Type: Download
Web Site: http://www.flightsimbrasil.comflytobrazil

Title: **Peru North & Lima 2002**
TecPilot Rating: ✈✈✈✈✈
Description: Fly 22 airports and 44 aerodromes, exploring the tourist and geographic marvels of Peru. Photorealistic scenery. Lima and Tarapoto mesh terrain recreated using DEM. Includes more than 100 buildings.
Pub/Dev/Source: Alfredo Mendiola Loyola
Rec Spec: Pentium III-1GHz or higher, 256 Mb or more RAM, 32 Mb 3D graphics card.
Platform: FS2002 *FS2004
Type: Download
Web Site: http://www.geocities.comperuscenery/PSEnglish.html

a-Z   FS Add-On - Scenery

Title:	**Philadelphia Intl (KPHL)**
TecPilot Rating:	✦✦✦✦✦
Description:	Includes accurate and detailed scenery of Philadelphia Intl KPHL including buildings, runways, taxiways and airport objects. Realistic buildings and objects with full texturing. Active docking system and jetways at each gate. serviceArmada(TM).
Pub/Dev/Source:	SimFlyers Associated
Rec Spec:	Pentium III-1GHz or higher, 256 Mb or more RAM, 32 Mb 3D graphics card.
Platform:	FS2000 FS2002 *FS2004
Type:	Download
Web Site:	http://www.simflyers.net

Title:	**Porto Santo 2002**
TecPilot Rating:	✦✦✦✦✦
Description:	The island lies 37 km (23 miles) to the Northeast of Madeira. On a clear day the islands can see each other on the horizon. A 9km (6 mile) stretch of natural sandy beach is the main attraction but the sleepy village and tranquil lifestyle is the main reason why visitors keep coming back.
Pub/Dev/Source:	Airsim/Jacques Sylvain
Rec Spec:	Pentium III-1GHz or higher, 256 Mb or more RAM, 32 Mb 3D graphics card.
Platform:	FS2002 *FS2004
Type:	Download
Web Site:	http://www.airsim.free.fr

Title:	**Real Azores**
TecPilot Rating:	✦✦✦✦✦
Description:	Located in the Atlantic, about 1500 Km west of Lisbon. All five major Airports (Flores, Horta, Lajes, Ponta Delgada and Santa Maria) included. Photorealistic buildings, runways with night effects. Accurate aeronautical data. Simple Terrain Mesh near airfields. Correct AI aircraft paths.
Pub/Dev/Source:	Predro Sousa
Rec Spec:	Pentium III-1GHz or higher, 256 Mb or more RAM, 32 Mb 3D graphics card.
Platform:	FS2002 *FS2004
Type:	Download
Web Site:	http://www.realazores.com

Title:	**Reggio Calabra Airport**
TecPilot Rating:	✦✦✦✦✦
Description:	Reggio Calabria and its airport are relatively unknown outside Italy. However, the location has some outstanding characteristics that make it special. First, its location is an ideal reference point for those who have a passion for IFR and VFR flights and who want to discover more about the Mediterranean. .
Pub/Dev/Source:	Pub/Dev/Source:
Rec Spec:	Pentium III-1GHz or higher, 256 Mb or more RAM, 32 Mb 3D graphics card.
Platform:	FS2002 *FS2004
Type:	Download
Web Site:	http://www.lagoonline.com

FS Add-On - Scenery   a-Z

Title: Reno & Lake Tahoe 10M Mesh & Landclass
TecPilot Rating: ✦✦✦✦✦
Description: 10 Meter Resolution Mesh Scenery and Landclass for the Reno and Lake Tahoe area. Designed with USGS 10m resolution DEM to offer the best level of detail. Mesh Designed with the USGS 1/3 elevation data and engineered to 9.6m resolution. Area of Coverage: N 40 W 121 - N38.75 W119.24.
Pub/Dev/Source: FSFreeware/Raimondo Taburet
Rec Spec: Pentium III-1GHz or higher, 256 Mb or more RAM, 32 Mb 3D graphics card.
Platform: FS2002 *FS2004
Type: Download
Web Site: http://www.fsfreeware.com

Title: Rhein-Ruhr 2002 Delux
TecPilot Rating: ✦✦✦✦✦
Description: Photorealistic terrain covering the area between Duisburg and Dortmund and many highly detailed sights in various cities. The terrain chunks include a full set of seasonal and night textures plus 3d objects. Also contains the accurately rendered airfields; EDLE and EDLM. .
Pub/Dev/Source: Reiffer Bros, Grueterich Software/Martin, Benedikt & Philipp Reiffer, Michael Grueterich
Rec Spec: Pentium III-1GHz or higher, 256 Mb or more RAM, 32 Mb 3D graphics card.
Platform: FS2000 (box) FS2002 (download)
Type: Download or Box
Web Site: http://www.flightsim.reiffer.com

Title: Rio Grande Do Norte & Sul
TecPilot Rating: ✦✦✦✦✦
Description: Contains six airports of the states of Rio Grande do Norte and Rio Grande do Sul. Included: Bagé (SBBG), Caxias do Sul (SBCX), Mossoró (SBMS), Natal (SBNT), Pelotas (SBPK) and Porto Alegre (SBPA).
Pub/Dev/Source: Flight Sim Brazil/Leandro Machado, Carlos Pereira
Rec Spec: Pentium III-1GHz or higher, 256 Mb or more RAM, 32 Mb 3D graphics card.
Platform: FS2002 *FS2004
Type: Download
Web Site: http://www.flightsimbrasil.comflytobrazil2

Title: Rome L.d.Vinci Intl (LIRF)
TecPilot Rating: ✦✦✦✦✦
Description: Includes Accurate and detailed scenery of Rome L.d.Vinci Intl LIRF including buildings, runways, taxiways and airport objects. Realistic buildings and objects with full texturing. Active docking system and jet ways at each gate. serviceArmada(TM).
Pub/Dev/Source: SimFlyers Associated
Rec Spec: Pentium III-1GHz or higher, 256 Mb or more RAM, 32 Mb 3D graphics card.
Platform: FS2000 FS2002 *FS2004
Type: Download
Web Site: http://www.simflyers.net

a-z

FS Add-On - Scenery

Title:	**San Francisco GMAX Works**
TecPilot Rating:	✦✦✦✦✦
Description:	Designed by FSFreeware in collaboration with Gmax Works. San Francisco blend, Auto-Gen scenery, custom designed Land Class, 10m mesh scenery. The scenery also includes night lights and comes with 10m mesh coverage specially adapted for the scenery which covers the area from Santa Rosa up north to Santa Cruz on the South. Good frames rates.
Pub/Dev/Source:	FSFreeware/Raimondo Taburet
Rec Spec:	Pentium III-1GHz or higher, 256 Mb or more RAM, 32 Mb 3D graphics card.
Platform:	FS2002 *FS2004
Type:	Download
Web Site:	http://www.fsfreeware.com

Title:	**Santander 2001/2**
TecPilot Rating:	✦✦✦✦✦
Description:	Comes with the airport, static aircraft and objects. Includes a high resolution mesh terrain of the approach areas through the mountains near Santander.
Pub/Dev/Source:	SimWings/Manfred Spatz
Rec Spec:	Pentium III-1GHz or higher, 256 Mb or more RAM, 32 Mb 3D graphics card.
Platform:	FS2000 FS2002 *FS2004
Type:	Download
Web Site:	http://www.sim-wings.de

Title:	**Scandinavia - Sweden, Norway, Denmark**
TecPilot Rating:	✦✦✦✦✦
Description:	Weighing in at 25Mb this mesh based scenery covers the areas as stated on the tin. Available via direct download or CD this will increase terrain detail by 40-50 meters in resolution. We recommend this for VFR aviators only. A demo is available at 426Kb.
Pub/Dev/Source:	FSFreeware/Raimondo Taburet
Rec Spec:	Pentium III-1GHz or higher, 256 Mb or more RAM, 32 Mb 3D graphics card.
Platform:	FS2000 FS2002 *FS2004
Type:	Box or Download
Web Site:	http://www.fsfreeware.com

Title:	**Scandinavia Land Class**
TecPilot Rating:	✦✦✦✦✦
Description:	Covers the whole area of Scandinavia including Denmark, Norway, Sweden, Finland and the Baltics. Land Class works with default or third party textures. Designed from real maps, the scenery draws all geographical features such as forests, fields, AutoGen scenery and cities. Exactly as they exist in in real life. Fully compatible with other Terrain Mesh Sceneries.
Pub/Dev/Source:	FSFreeware/Raimondo Taburet
Rec Spec:	Pentium III-1GHz or higher, 256 Mb or more RAM, 32 Mb 3D graphics card.
Platform:	FS2002 *FS2004
Type:	Download
Web Site:	http://www.fsfreeware.com

The Good Flight Simmer's Guide Mk.II - 2003

FS Add-On - Scenery　　　　　　　　　　a-Z

Title: **Scandinavian Airports**
TecPilot Rating: ✦✦✦✦✦
Description: Discover 8 Scandinavian airports. Every airport is highly detailed recreated with many objects. Stunning night effects, dynamic ground traffic and the large variety of typical airport elements form an impressive simulation. Dynamic scenery is compatible with default AI-traffic. A new detailed Mesh-scenery for some of the airports completes the product.
Pub/Dev/Source: Aerosoft
Rec Spec: Pentium III-1GHz or higher, 256 Mb or more RAM, 32 Mb 3D graphics card.
Platform: FS2002 *FS2004
Type: Box
Web Site: http://www.aerosoft.com

Title: **Scenery Airport Dübendorf**
TecPilot Rating: ✦✦✦✦✦
Description: A highly detailed rendition of Duebendorf airport which includes popular buildings like the museum, JU-hall, shelters, tower and further halls. Static aircraft ILS approach aids featured. Night lightning. Dynamic Scenery: Starting and landing jets: two Mirage approach during a Tiger taxi from the shelter to the runway. Driving tank lorries and fire engines. A MD-11 coming from Zuerich-Kloten is flying over Duebendorf.
Pub/Dev/Source: Aerosoft
Rec Spec: Pentium III-1GHz or higher, 256 Mb or more RAM, 32 Mb 3D graphics card.
Platform: FS2002 *FS2004
Type: Box
Web Site: http://www.aerosoft.com

Title: **Scenery Balearic Islands (Spain 1)**
TecPilot Rating: ✦✦✦✦✦
Description: Includes all the five airports of the Balearic Islands. Many objects and details have been lovingly recreated. Even small airfields such as Son Bonet and Aeroclub de Mahon have been reconstructed just as the originals were built.
Pub/Dev/Source: Aerosoft/SimWings
Rec Spec: Pentium III-1GHz or higher, 256 Mb or more RAM, 32 Mb 3D graphics card.
Platform: FS2000 FS2002 *FS2004
Type: Box
Web Site: http://www.aerosoft.com

Title: **Scenery Budapest 2002**
TecPilot Rating: ✦✦✦✦
Description: The Hungarian capital and a area surrounding it. Photorealistic coverage, with a relatively high pixel resolutions of 4m/p. The mesh is divided into approximately 150 meters per altitude point. It consists of five airports in the city namely: Ferihegy Int., Budaörs, Farkashegy, Harmashatarhegy and Tököl (millitary).
Pub/Dev/Source: Aerosoft/Andras Kozma
Rec Spec: Pentium III-1GHz or higher, 256 Mb or more RAM, 32 Mb 3D graphics card.
Platform: FS2002 *FS2004
Type: Box
Web Site: http://www.aerosoft.com

a-Z   FS Add-On - Scenery

Title:	**Scenery Budapest Millennium Edition 2002 (Online Version)**
TecPilot Rating:	✦✦✦✦✦
Description:	The Hungarian capital and a area surrounding it. Photorealistic coverage, with a relatively high pixel resolutions of 4m/p. The mesh is divided into approximately 150 meters per altitude point. It consists of five airports in the city namely: Ferihegy Int., Budaörs, Farkashegy, Harmashatarhegy and Tököl (millitary)
Pub/Dev/Source:	Vistamare Software/Andras Kozma
Rec Spec:	Pentium III-1GHz or higher, 256 Mb or more RAM, 32 Mb 3D graphics card.
Platform:	FS2002 *FS2004
Type:	Download
Web Site:	http://www.vistamaresoft.com

Title:	**Scenery Canary Islands**
TecPilot Rating:	✦✦✦✦✦
Description:	Islas Canarias. Adds 9 outstanding airports and new landscapes. Tenerife (main island of the Canaries, with 2 airports), Gran Canaria (most famous island with 1 airport and 1 airfield), Lanzarote, (1 airport), Fuerteventure (1 airport), La Gomera (1 new airport), La Palma (1 airport), El Hierro (smallest of the Canary Islands with a new airport).
Pub/Dev/Source:	Aerosoft/SimWings
Rec Spec:	Pentium III-1GHz or higher, 256 Mb or more RAM, 32 Mb 3D graphics card.
Platform:	FS2002 *FS2004
Type:	Box
Web Site:	http://www.aerosoft.com

Title:	**Scenery Greece**
TecPilot Rating:	✦✦✦✦✦
Description:	Greece with it's islands and the 25 largest airports. Many detailed objects as well as satellite based VFR scenery Sourced from GEOSYS Data. English/French manuals included.
Pub/Dev/Source:	Aerosoft
Rec Spec:	Pentium III-1GHz or higher, 256 Mb or more RAM, 32 Mb 3D graphics card.
Platform:	FS2002 *FS2004
Type:	Box
Web Site:	http://www.aerosoft.com

Title:	**Scenery Honolulu 2002**
TecPilot Rating:	✦✦✦✦✦
Description:	Includes the international airport set in an exotic (or was that erotic?) back drop. The walk ways from the main terminal are lined with palm trees and are open, allowing the trade winds to waft through. (Puke, how romantic, Elvis!) In the terminal there are occasional Hawaiian performances of music and hula dancing. (Yawn!)
Pub/Dev/Source:	Aerosoft
Rec Spec:	Pentium III-1GHz or higher, 256 Mb or more RAM, 32 Mb 3D graphics card.
Platform:	FS2002 *FS2004
Type:	Box
Web Site:	http://www.aerosoft.com

FS Add-On - Scenery   a-z

Title: **Scenery Spanish Airports (Spain 2)**
TecPilot Rating: ✦✦✦✦✦
Description: Add 5 detailed airports and characteristic objects. Developed by the same team who designed Spain 1 (Balearic Islands). It adds five further Spanish airports and surrounding landscapes. Alicante, Barcelona, Valencia, Pamplona and Santander.
Pub/Dev/Source: Aerosoft/SimWings
Rec Spec: Pentium III-1GHz or higher, 256 Mb or more RAM, 32 Mb 3D graphics card.
Platform: FS2002 *FS2004
Type: Box
Web Site: http://www.aerosoft.com

Title: **Sky Ranch**
TecPilot Rating: ✦✦✦✦✦
Description: See the world from a different perspective. Grassy Meadows. Sky Ranch (UT47) is located in southern Utah. It maximizes the full potential of FS2002's scenery engine. A little niche but someone will love using it as a base to explore the wonderful region.
Pub/Dev/Source: Abacus
Rec Spec: Pentium III-1GHz or higher, 256 Mb or more RAM, 32 Mb 3D graphics card.
Platform: FS2002
Type: Box
Web Site: http://www.abacuspub.com

Title: **South Africa**
TecPilot Rating: ✦✦✦✦✦
Description: A high-detail terrain mesh scenery designed at 1 Arc/Sec resolution (very high resolution). Covers the entire region of South Africa. We recommend this for VFR aviators only. A demo is available at 537Kb.
Pub/Dev/Source: FSFreeware/Raimondo Taburet
Rec Spec: Pentium III-1GHz or higher, 256 Mb or more RAM, 32 Mb 3D graphics card.
Platform: FS2000 FS2002 *FS2004
Type: Box or Download
Web Site: http://www.fsfreeware.com

Title: **South East England - Part 3**
TecPilot Rating: ✦✦✦✦✦
Description: A collection of 24 Airports in the southeast of England made to a remarkable quality and realism. This product also includes "London Gatwick" Airport Scenery. All buildings accurately modelled and textured. Taxiway/stand markings are realistic. BRITISH runway markings. Very detailed holding points with guard lights (Wig Wags). Sheep! 3D Trees. Interactive objects. 'Mike' the Marshaller.
Pub/Dev/Source: UK2000 Scenery/Gary Summons
Rec Spec: Pentium III-1GHz or higher, 256 Mb or more RAM, 32 Mb 3D graphics card.
Platform: FS2000 FS2002 *FS2004
Type: Download
Web Site: http://www.uk2000scenery.com

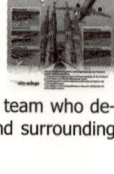

a-Z FS Add-On - Scenery

Title:	**South West England - Part 2**
TecPilot Rating:	✦✦✦✦✦
Description:	A collection of 25 Airports in the south west of England made to remarkable quality and realism. All buildings accurately modelled and textured. Taxiway/stand markings that are realistic. BRITISH runway markings. Very detailed holding points with guard lights (Wig Wags). Sheep! 3D Trees. Interactive objects. 'Mike' the Marshaller.
Pub/Dev/Source:	UK2000 Scenery/Gary Summons
Rec Spec:	Pentium III-1GHz or higher, 256 Mb or more RAM, 32 Mb 3D graphics card.
Platform:	FS2000 FS2002 *FS2004
Type:	Download
Web Site:	http://www.uk2000scenery.com

Title:	**Spain, Portugal, Morocco, Canary Islands**
TecPilot Rating:	✦✦✦✦✦
Description:	A high-detail terrain mesh scenery designed at 1 Arc/Sec resolution. Covers the entire regions of Spain, Portugal, Morocco and the Canary Islands. We recommend this for VFR aviators only. A demo is available at 803Kb.
Pub/Dev/Source:	FSFreeware/Raimondo Taburet
Rec Spec:	Pentium III-1GHz or higher, 256 Mb or more RAM, 32 Mb 3D graphics card.
Platform:	FS2000 FS2002 *FS2004
Type:	Box or Download
Web Site:	http://www.fsfreeware.com

Title:	**St. Petersburg 2002**
TecPilot Rating:	✦✦✦✦✦
Description:	Includes Pulkovo airport. Custom buildings, apron and ground textures. Weathered apron effects with oil slicks. Detailed taxiway, runway, and apron markings. Many supported buildings, trees, hedges and grasses. Vehicles on aprons, including GPU, airport tankers, vans and maintenance.
Pub/Dev/Source:	Sergay Carr
Rec Spec:	Pentium III-1GHz or higher, 256 Mb or more RAM, 32 Mb 3D graphics card.
Platform:	FS2002 *FS2004
Type:	Download
Web Site:	http://www.strp.fsnet.co.uk

Title:	**Stockholm Arlanda (ESSA)**
TecPilot Rating:	✦✦✦✦✦
Description:	Fully seasonal and night ground texturing. Most buildings are exquisitely photo textured; now in 16 bit Hi-res and night lighted. Current CAA high scale drawings used for the airport layouts. Latest navigational information and databases used. Full real-life SAFEGATE®, APIS++® and APIS® Docking Systems at all relevant gates. Instruction file in PDF-format included. .
Pub/Dev/Source:	SwedFlight Pro
Rec Spec:	Pentium III-1GHz or higher, 256 Mb or more RAM, 32 Mb 3D graphics card.
Platform:	FS2002 *FS2004
Type:	Download
Web Site:	http://www.swedflight.com

FS Add-On - Scenery   a-Z

Title: Stockholm Arlanda Intl
TecPilot Rating: ✦✦✦✦✦
Description: Realistically rendered airport with lots of nicely textured buildings. Photographs from the real airport were used as a reference. AI compatible. Static aircraft and vehicles. Dynamic scenery. Fully functional Wig-wags (Scandinavian type). Landclass for the region and Manual in PDF format.
Pub/Dev/Source: FS Dream Factory/Cornel Grigoriu
Rec Spec: Pentium III-1GHz or higher, 256 Mb or more RAM, 32 Mb 3D graphics card.
Platform: FS2002 *FS2004
Type: Download
Web Site: http://www.fsdreamfactory.com

Title: Sturup 2002
TecPilot Rating: ✦✦✦✦✦
Description: AI compatible. Realistically rendered airport with a lots of details including buildings with photorealistic textures. Dynamic and static scenery such as catering trucks, pushback cars, ladders etc. Slightly meshed terrain around the airport (the area is flat but it has some hills). Beautifully rendered gates with transparent windows and a lots of other surprises. .
Pub/Dev/Source: FS Dream Factory/Cornel Grigoriu
Rec Spec: Pentium III-1GHz or higher, 256 Mb or more RAM, 32 Mb 3D graphics card.
Platform: FS2002 *FS2004
Type: Download
Web Site: http://www.fsdreamfactory.com

Title: SwedFlight Pro
TecPilot Rating: ✦✦✦✦✦
Description: The SwedFlight scenery is a collection of three Swedish airports, Stockholm, Malmoe, and Gothenborg. The airports feature seasonal and night effects, airport buildings and unique textures created from over 400 pictures. All taxiways are signed and lighted, they are also drawn according to various navigational diagrams.
Pub/Dev/Source: SwedFlight
Rec Spec: Pentium 600, 256 MByte RAM, 3D Graphics Card
Platform: FS2002
Type: Payware
Web Site: http://www.swedflight.com

Title: Swiss Airports
TecPilot Rating: ✦✦✦✦✦
Description: 7 famous Swiss airports in a high quality. Includes. Zurich, Geneva, Bern, Lugano, Sion, Basel/Mulhouse, Altenrhein (Lake Constance). <u>Not released at time of publication</u>.
Pub/Dev/Source: Aerosoft
Rec Spec: Pentium III-1GHz or higher, 256 Mb or more RAM, 32 Mb 3D graphics card.
Platform: FS2002 *FS2004
Type: Box
Web Site: http://www.aerosoft.com

a-Z　　　　　　　　　　　　　　FS Add-On - Scenery

Title:	**Switzerland 3**
TecPilot Rating:	✦✦✦✦✦
Description:	Photorealistic elevation model of Switzerland. Raster elevation model snapshot at 200 Meters. Resolution photos taken at 12 Meters. All airports included. Very little information provided in English so stick your finger in the jam and find by getting it!
Pub/Dev/Source:	Flight & Cockpit/Mice
Rec Spec:	Pentium III-1GHz or higher, 256 Mb or more RAM, 32 Mb 3D graphics card.
Platform:	FS2000 FS2002 *FS2004
Type:	Box
Web Site:	http://www.pcflight.ch

Title:	**Taipei, Taiwan**
TecPilot Rating:	✦✦✦✦✦
Description:	Realistic terminal buildings, Taxi lines, trees, Taiwan style houses around the airport, CKS airport and Sung Shan Airports, ships in the sea, Taipei "Memorial Hall", Grand Hotel and lots more
Pub/Dev/Source:	Samsoft
Rec Spec:	Pentium 600, 64 MByte RAM, 3D Graphics Card.
Platform:	FS2000 FS2002
Type:	Download
Web Site:	http://www.samsoftpub.com

Title:	**Tenerife & LA Gomera 2003**
TecPilot Rating:	✦✦✦✦✦
Description:	Contains high resolution mesh terrain of the islands. Also features detailed airports and landclass files to (partly) change texturing. AI-traffic compatible. Static aircraft, removable for AI-Traffic use. All textures made from photographs taken at real airports. Night effects.
Pub/Dev/Source:	SimWings/Thorsten Loth
Rec Spec:	Pentium III-1GHz or higher, 256 Mb or more RAM, 32 Mb 3D graphics card.
Platform:	FS2002 *FS2004
Type:	Download
Web Site:	http://www.sim-wings.de

Title:	**Terrain Puerto Rico - Virgin Islands**
TecPilot Rating:	✦✦✦✦✦
Description:	USGS Ned dem 38.2m Mesh Terrain Puerto Rico - Virgin Island. Designed with the Latest Dem Data from Usgs. Offers a very detailed Terrain Mesh for Puerto Rico and the Virgin Islands.
Pub/Dev/Source:	FSFreeware/Raimondo Taburet
Rec Spec:	Pentium III-1GHz or higher, 256 Mb or more RAM, 32 Mb 3D graphics card.
Platform:	FS2002 *FS2004
Type:	Download
Web Site:	http://www.fsfreeware.com

The Good Flight Simmer's Guide Mk.II - 2003

FS Add-On - Scenery   a-Z

Title: **Terramesh Europa 2002**
TecPilot Rating: ✈ ✈ ✈ ✈ ✈
Description: Upgrade the terrain of Flight Simulator with Lago's replacement European Terrain Mesh! Hills, mountains and valleys just as they should be - provides a remarkable improvement to all your low level flights. In one boxed edition on two CD-ROMs; provides coverage for the UK, Ireland, France, Spain, Portugal, Germany, Austria, Switzerland, Italy, Belgium, Holland, Denmark, Sweden and Norway.
Pub/Dev/Source: Just Flight/Lago
Rec Spec: Pentium III-1GHz or higher, 256 Mb or more RAM, 32 Mb 3D graphics card.
Platform: FS2002 *FS2004
Type: Box
Web Site: http://www.justflight.com

Title: **The Midlands - Part 5**
TecPilot Rating: ✈ ✈ ✈ ✈ ✈
Description: 42 Fully Detailed Airfields. 300MB of purpose build models. The Area covered is the largest in the UK2000 Scenery project. Starting in the southwest corner we have the old RAF Kemble and going up to northeast corner at Nottingham, many famous airfields are included like Cranfield, Coventry and Old Warden.
Pub/Dev/Source: UK2000 Scenery/Gary Summons
Rec Spec: Pentium III-1GHz or higher, 256 Mb or more RAM, 32 Mb 3D graphics card.
Platform: FS2000 FS2002 *FS2004
Type: Download
Web Site: http://www.uk2000scenery.com

Title: **The Rockies 38m Terrain**
TecPilot Rating: ✈ ✈ ✈ ✈ ✈
Description: From the western edge of the Great Plains to the Great Basin and the Sonoran Desert, from New Mexico to the Canadian border, the entire Rocky Mountain System is brought to life in 38.3m terrain resolution. Pikes Peak, Mt Elbert, The Spanish Peaks, Boulder Pass, it's all here - along with The Great American Southwest. Re-sampled from the USGS 30m National Elevation Database, detail and accuracy is ensured. .
Pub/Dev/Source: FSGenesis
Rec Spec: Pentium III-1GHz or higher, 256 Mb or more RAM, 32 Mb 3D graphics card.
Platform: FS2002 *FS2004
Type: Box
Web Site: http://www.fsgenesis.com

Title: **Toronto Pearson Intl (CYYZ)**
TecPilot Rating: ✈ ✈ ✈ ✈
Description: Accurate and detailed scenery of Toronto Pearson Intl Airport (CYYZ) including buildings, runways, taxiways and airport objects. Realistic buildings and objects with full texturing.
Pub/Dev/Source: SimFlyers Associated
Rec Spec: Pentium III-1GHz or higher, 256 Mb or more RAM, 32 Mb 3D graphics card.
Platform: FS2002 *FS2004
Type: Download
Web Site: http://www.simflyers.net

a-Z   FS Add-On - Scenery

Title:	**United Kingdom**
TecPilot Rating:	✢ ✢ ✢ ✢ ✢
Description:	Hi resolution mesh scenery designed at 1 Arc/Sec resolution. Covers England, Scotland, Northern Ireland, Southern Ireland, Wales and the Channel Islands. We recommend this for VFR aviators only. A demo is available at 549Kb.
Pub/Dev/Source:	FSFreeware/Raimondo Taburet
Rec Spec:	Pentium III-1GHz or higher, 256 Mb or more RAM, 32 Mb 3D graphics card.
Platform:	FS2000 FS2002 *FS2004
Type:	Box or Download
Web Site:	http://www.fsfreeware.com

Title:	**VFR Photographic Scenery: Central & Southern England**
TecPilot Rating:	✢ ✢ ✢ ✢ ✢
Description:	Made by draping high quality aerial photography over a 3D model of flight simulators terrain. This Central and Southern region package starts at the south coast of England and extends north to a line at N53º14.4´ running east-west just north of Chester. The eastern boundary is along a line at W0º40.2´ running north-south from Sandsend (just west of Whitby, North Yorks) to Bognor Regis. The western boundary is along a line at W3º18.6´ from the mouth of the River Dee to the Estuary of the River Otter, east of Budleigh Salterton.
Pub/Dev/Source:	John Farrie (Visual Flight), GetMapping PLC/Just Flight
Rec Spec:	Pentium III-1GHz or higher, 256 Mb or more RAM, 32 Mb 3D graphics card.
Platform:	FS2002 *FS2004
Type:	Box
Web Site:	http://www.justflight.com

Title:	**VFR Photographic Scenery: East and South-East England**
TecPilot Rating:	✢ ✢ ✢ ✢ ✢
Description:	You can now experience true VFR flying, with photographic quality detail including roads, fields, towns and villages as well as hills, valleys and airport approaches... exactly as would be seen from a real aircraft. Fly over your street, school, office, local football ground and any other visible landmarks! Made by draping high quality aerial photography over a 3D model of flight simulators terrain. This first volume, with approximately 1.8Gb of data, comes on 3 CD-ROMs and covers the area of England east of a line at W0º48.34' running north-south from Staithes (near Whitby, North Yorks) to Selsey (near Chichester, East Sussex).
Pub/Dev/Source:	John Farrie (Visual Flight), GetMapping PLC/Just Flight
Rec Spec:	Pentium III-1GHz or higher, 256 Mb or more RAM, 32 Mb 3D graphics card.
Platform:	FS2002 *FS2004
Type:	Box
Web Site:	http://www.justflight.com

FS Add-On - Scenery   a-Z

Title:	**VFR Photographic Scenery: Wales & South-West England**
TecPilot Rating:	✦✦✦✦✦
Description:	This Wales and South-West England region package starts at the south coast of Devon and Cornwall, running up to the north coast of Wales. The eastern boundary is along a line at W2º46.48´ running north-south through Wrexham.
Pub/Dev/Source:	John Farrie (Visual Flight), GetMapping PLC/Just Flight
Rec Spec:	Pentium III-1GHz or higher, 256 Mb or more RAM, 32 Mb 3D graphics card.
Platform:	FS2002 *FS2004
Type:	Box
Web Site:	http://www.justflight.com

Title:	**VFR-Airfields Vol. 1**
TecPilot Rating:	✦✦✦✦✦
Description:	The first in a sequel further planned to cover the whole of Germany with select airfields for VFR flights. With the further perfection of elevation and texture depiction, VFR flying has become the most rewarding way to enjoy simulated flight. You may actually recognise the area you are in and identify VFR relevant data. To further enhance the quality of the area covered, VFR-Airfields include FS-LandClass enhancement of the area covered.
Pub/Dev/Source:	Aerosoft
Rec Spec:	Pentium III-1GHz or higher, 256 Mb or more RAM, 32 Mb 3D graphics card.
Platform:	FS2002 *FS2004
Type:	Box
Web Site:	http://www.aerosoft.com

Title:	**Valencia 2002**
TecPilot Rating:	✦✦✦✦✦
Description:	Features a detailed version of Valencia International airport, which includes the old "Manisa" part. Contains a special texture order (Landclass) at Valencia to display the size of the city and port of Valencia. The airport is AI compatible.
Pub/Dev/Source:	SimWings/Manfred Spatz
Rec Spec:	Pentium III-1GHz or higher, 256 Mb or more RAM, 32 Mb 3D graphics card.
Platform:	FS2002 *FS2004
Type:	Download
Web Site:	http://www.sim-wings.de

2500 members from over 60 Countries use TecPilot weekly.
Over 1 million people visit TecPilot for their sim news every year!

WWW.TECPILOT.COM
Home Of "The Good Flight Simmer's Guide"

a-Z   FS Add-On - Scenery

Title: **Washington Dulles (KIAD)**
TecPilot Rating: ✦✦✦✦✦
Description: Includes detailed scenery of Washington Dulles KIAD including buildings, runways, taxiways and airport objects. Realistic buildings and objects with full texturing. Active docking system and jet ways at each gate. serviceArmada(TM).
Pub/Dev/Source: SimFlyers Associated
Rec Spec: Pentium III-1GHz or higher, 256 Mb or more RAM, 32 Mb 3D graphics card.
Platform: FS2000 FS2002 *FS2004
Type: Download
Web Site: http://www.simflyers.net

Title: **West Coast 38m Terrain**
TecPilot Rating: ✦✦✦✦✦
Description: From the Northern Cascades to the Bay Area, from San Fernando Valley to Death Valley to San Diego--and all points in between, you'll fly over the most accurate terrain possible given current processing power. Re-sampled to 38.2m resolution from USGS 30m source files to minimize the error factor.
Pub/Dev/Source: FSGenesis
Rec Spec: Pentium III-1GHz or higher, 256 Mb or more RAM, 32 Mb 3D graphics card.
Platform: FS2002 *FS2004
Type: Box
Web Site: http://www.fsgenesis.com

Title: **Wonderful Rio (Expansion)**
TecPilot Rating: ✦✦✦✦✦
Description: 6 new aerodromes added. Photorealistic terrain area increased covering from Guanabara Bay to Marambaia. Autogen of over 3200 sq. miles was designed to populate the scenery, increasing VFR flights substantially. SBJR - Jacarepaguá Airport, SBAF - Campo dos Afonsos Air Base, SBSC - Santa Cruz Air Base, SDMC - Maricá Airport, SDIN - Clube CEU, SDMR - Marambaia Military Airfield. AFCAD™ files included.
Pub/Dev/Source: RealFlight
Rec Spec: Pentium III-1GHz or higher, 256 Mb or more RAM, 32 Mb 3D graphics card.
Platform: FS2002 *FS2004
Type: Download
Web Site: http://www.realflight.com.br

Title: **Wonderful Rio 2002**
TecPilot Rating: ✦✦✦✦✦
Description: Includes: 2 main airports (Santos-Dumont and Galeão International). 5 Helipads around the city. 1600sq miles of photo-terrain textures in an ortophotographic resolution. 3D objects and landmarks all around the city and all built in Gmax™ technology. Enhanced traffic files for use with TTOOLS™. AFCAD™ Included. Compatible with MyTraffic™. Lastly a Bonus Gmax™ Varig Log Boeing 727-200F and Embraer Regional Jet 145.
Pub/Dev/Source: RealFlight
Rec Spec: Pentium III-1GHz or higher, 256 Mb or more RAM, 32 Mb 3D graphics card.
Platform: FS2002 *FS2004
Type: Download
Web Site: http://www.realflight.com.br

FS Add-On - Scenery   a-Z

Title: **World Airports**
TecPilot Rating: ✦✦✦✦✦
Description: The USA - one of the world's busiest air traffic regions and widely neglected by scenery designers. Now nine of the most detailed sceneries for FS2000 & 2002, featuring 6 of the biggest and busiest airports in the US and worldwide, plus 3 popular European destinations are available. Includes Atlanta, Bari, Dallas Fort Worth, Los Angeles, Manchester, Newark, Orlando, Rome, Philadelphia.
Pub/Dev/Source: SimFlyers/Just Flight
Rec Spec: Pentium III-1GHz or higher, 256 Mb or more RAM, 32 Mb 3D graphics card.
Platform: FS2000 FS2002 *FS2004
Type: Box
Web Site: http://www.justflight.com

Title: **Yellowstone & Grand Tetons**
TecPilot Rating: ✦✦✦✦✦
Description: Yellowstone & Grand Tetons National Parks in their entirety, including a significant surrounding area. Compiled from USGS 7.5 minute 10-meter digital elevation models. The United States Geological Survey, EROS Data Center, Sioux Falls, SD. 10m 7.5 minute dataset.
Pub/Dev/Source: FSGenesis
Rec Spec: Pentium III-1GHz or higher, 256 Mb or more RAM, 32 Mb 3D graphics card.
Platform: FS2002 *FS2004
Type: Box
Web Site: http://www.fsgenesis.com

Title: **Zurich Kloten Airport**
TecPilot Rating: ✦✦✦✦✦
Description: Designed for FS2002, SwissVFR or Switzerland3. Airport layout plan of 2006. Gangways at all gates, docking system, all runways and navaids, taxiways with center lines and signs, parking facilities, freight buildings and hangars, helicopter base, night lighting apron and static planes at the finger docks.
Pub/Dev/Source: MICE Innovative Computer Engineering/Hans-Jörg Müller
Rec Spec: Pentium III-1GHz or higher, 256 Mb or more RAM, 32 Mb 3D graphics card.
Platform: FS2000 (?) FS2002 *FS2004
Type: Download
Web Site: http://www.mice.ch

For the very latest downloads including
scenery, aircraft and utilities visit...

WWW.SIMMARKET.COM
Your one stop flight shop!

a-Z
FS Add-On
Utilities

FS Add-On - Utilities   a-Z

Title: **ActiveSky wxRE**
TecPilot Rating: ✦✦✦✦✦
Description: A next-generation weather environment engine for flight simulator. Realistic modelling of dynamic changing weather according to real-world behaviour - available in either LIVE or Offline/Preset mode. In LIVE mode, your weather environment will be a replica of real-world/time conditions. In offline mode you can use downloaded CYCLE files or make weather settings. Either way, ActiveSky wxRE can dynamically change the weather according to live or preset forecast reports (TAFs)! .
Pub/Dev/Source: HiFi Simulation Software/Damian Clark
Rec Spec: Pentium III-1GHz or higher, 256 Mb or more RAM, 32 Mb 3D graphics card.
Platform: FS2002 *FS2004
Type: Download
Web Site: http://www.hifi.avsim.net/activesky

Title: **Aircraft Animator**
TecPilot Rating: ✦✦✦✦✦
Description: AA was quite a treat to aircraft developer's ears when it arrived. The program, allows even the less than average aircraft developer to add dynamic surfaces to a plane (gear, flaps, spoilers, propeller, and shining landing lights).
Pub/Dev/Source: Abacus
Rec Spec: Pentium III-1GHz or higher, 256 Mb or more RAM, 32 Mb 3D graphics card.
Platform: FS98 FS2000 (*FS2002) CFS
Type: Box
Web Site: http://www.abacuspub.com

Title: **Aircraft Container Manager**
TecPilot Rating: ✦✦✦✦✦
Description: A tool that allows you to organize aircraft in Flight Simulator. Features include the ability to organize aircraft folders, change data in **.cfg** files, view aircraft with textures, view and export aircraft textures and change station loads.
Pub/Dev/Source: DVData
Rec Spec: Pentium III-1GHz or higher, 256 Mb or more RAM, 32 Mb 3D graphics card.
Platform: FS2000 FS2002 CFS2 CFS3
Type: Download
Web Site: http://www.aircraftmanager.com

Title: **Aircraft Sound Manager**
TecPilot Rating: ✦✦✦✦✦
Description: ASM is a tool that allows you to organize aircraft sounds in Flight Simulator. There is a evaluation version of ASM available to demo most of the features.
Pub/Dev/Source: DVData
Rec Spec: Pentium III-1GHz or higher, 256 Mb or more RAM, 32 Mb 3D graphics card.
Platform: FS2000 FS2002 CFS2 CFS3 FS-ACOF
Type: Download
Web Site: http://www.aircraftmanager.com

a-Z FS Add-On - Utilities

Title:	**Airliners Env 2002**
TecPilot Rating:	✈ ✈ ✈ ✈
Description:	"Flight Crew On Board" & dynamic weather generator. Create a real commercial flight environment on all your flights with this dynamic interactive utility. More than 150 supported sound and weather events.
Pub/Dev/Source:	Hendrik Kupfernagel
Rec Spec:	Pentium III-1GHz or higher, 256 Mb or more RAM, 32 Mb 3D graphics card.
Platform:	FS2002 *FS2004
Type:	Download
Web Site:	http://www.simmarket.com

Title:	**Airport & Scenery Designer v2.1**
TecPilot Rating:	✈ ✈ ✈ ✈
Description:	For advanced and professionals designers, Airport & Scenery Designer's enormous capabilities lets you choose the right tool for adding the perfect scenery element. Whether it's runways, approach lighting, roads, rivers or mountains, Airport & Scenery Designer can handle it. You might say: "There's no more powerful software in the world for creating your flight sim world".
Pub/Dev/Source:	Abacus
Rec Spec:	Pentium III-1GHz or higher, 256 Mb or more RAM, 32 Mb 3D graphics card.
Platform:	FS2002 *FS2004 CFS2
Type:	Box
Web Site:	http://www.abacuspub.com

Title:	**ChartsViews2**
TecPilot Rating:	✈ ✈ ✈ ✈ ✈
Description:	ChartViews2 is a 'stand-alone' Windows utility for any Windows based flight sim that allows a number of airport charts to be available on screen while still flying and being able to have full access to a simulator's functions.
Pub/Dev/Source:	XHYTEX/Alex Hyett
Rec Spec:	Pentium III-1GHz or higher, 256 Mb or more RAM, 32 Mb 3D graphics card.
Platform:	ALL
Type:	Download
Web Site:	http://www.xhytex.comchartviews

Title:	**Custom Panel Designer**
TecPilot Rating:	✈ ✈ ✈ ✈ ✈
Description:	Custom Panel Designer, is a simple to use Flight Sim panel designer. After loading a bitmap panel image, you simply drag, and size gauges into there appropriate positions.
Pub/Dev/Source:	Abacus
Rec Spec:	Pentium III-1GHz or higher, 256 Mb or more RAM, 32 Mb 3D graphics card.
Platform:	FS98 FS2000 FS2002 CFS CFS2
Type:	Box
Web Site:	http://www.abacuspub.com

FS Add-On - Utilities   a-Z

Title: **EZ-Landmark**
TecPilot Rating: ✦✦✦✦✦
Description: EZ-Landmark eliminates the problem of guessing what stadium is which, or what's the name of the airport you're flying above. The three step installation places a name tag of sorts over more than 1000 landmarks in metropolitan areas in MSFS.
Pub/Dev/Source: Abacus
Rec Spec: Pentium III-1GHz or higher, 256 Mb or more RAM, 32 Mb 3D graphics card.
Platform: FS2000 FS2002
Type: Payware
Web Site: http://www.abacuspub.com

Title: **FS Action Scenery**
TecPilot Rating: ✦✦✦✦✦
Description: Using FS Action! Scenery you add this missing element to all your flights. In just a few minutes you can animate any of your "static scenery areas" with this easy-to-use tool. Designer Louis Sinclair makes it easy for you to add dynamic scenery everywhere.
Pub/Dev/Source: Abacus
Rec Spec: Pentium III-1GHz or higher, 256 Mb or more RAM, 32 Mb 3D graphics card.
Platform: FS98 FS2000
Type: Box
Web Site: http://www.abacuspub.com

Title: **FS Assist 2002**
TecPilot Rating: ✦✦✦✦✦
Description: Includes a "PushBack" module that will provide several options. Unlike the basic pushback option in FS this one includes a visual system, truck and Mat the pushback kid who will assist you. A "Framerate Wizard" - but better than in FS2000/02. Also includes and AutoSave facility and Key Assigner for Lago specific add-ons.
Pub/Dev/Source: Lago
Rec Spec: Pentium III-1GHz or higher, 256 Mb or more RAM, 32 Mb 3D graphics card.
Platform: FS2000 FS2002 *FS2004
Type: Download
Web Site: http://www.lagoonline.com

Title: **FS Clouds 2000**
TecPilot Rating: ✦✦✦✦✦
Description: Fly through practically unlimited skyscapes using the more than 60 new cloud components. See new dimensions in the sky as you fly past clouds that appear much more vaporous and volumetric. .
Pub/Dev/Source: Flight 1
Rec Spec: Pentium III-1GHz or higher, 256 Mb or more RAM, 32 Mb 3D graphics card.
Platform: FS2000 FS2002 *FS2004
Type: Box
Web Site: http://www.justflight.com

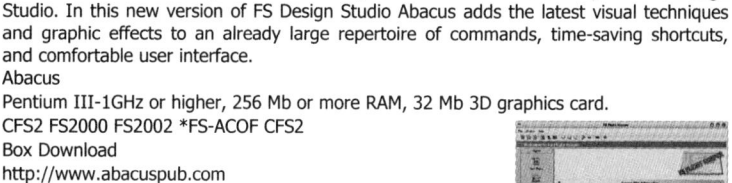

a-Z　　　　　　　　　　　　　　　　FS Add-On - Utilities

Title: **FS Design Studio v2**
TecPilot Rating: ✦✦✦✦✦
Description: FS Design Studio is "Flight Sim friendly" and powerful too. This easy to use software gives you the tool for making professional quality aircraft and 3D scenery. Of the tens of thousands of aircraft available, the overwhelming number of them were made with FS Design Studio. In this new version of FS Design Studio Abacus adds the latest visual techniques and graphic effects to an already large repertoire of commands, time-saving shortcuts, and comfortable user interface.
Pub/Dev/Source: Abacus
Rec Spec: Pentium III-1GHz or higher, 256 Mb or more RAM, 32 Mb 3D graphics card.
Platform: CFS2 FS2000 FS2002 *FS-ACOF CFS2
Type: Box Download
Web Site: http://www.abacuspub.com

Title: **FS Flight Keeper**
TecPilot Rating: ✦✦✦✦✦
Description: FS Flight Keeper (FSFK) combines two programs in one: A Logbook and a basic Aircraft Black Box. The Logbook keeps track of the flights you mike. It stores flight times (also day and night), Fuel used, pilot, aircraft used, etc. While the Black Box logs all aircraft events (e.g. autopilot settings, engine settings, weather, etc) FSFK also features flight weather planning, data export, flight critique and more.
Pub/Dev/Source: Thomas Molitor
Rec Spec: Pentium III-1GHz or higher, 256 Mb or more RAM, 32 Mb 3D graphics card.
Platform: FS2002 *FS-ACOF
Type: Download
Web Site: http://www.molitor-home.de/fs

Title: **FS Flight Log**
TecPilot Rating: ✦✦✦✦✦
Description: FS Flight Log is not just an ordinary logbook utility for Flight Simulator it actually records flight information from your active flight automatically! Make manual entries without using ACARS. Drop down list of your FS aircraft. Drop down list of ICAO Airport codes and names. Create custom HTML Log Reports.
Pub/Dev/Source: JordSoft/Keith Lovell
Rec Spec: Pentium III-1GHz or higher, 256 Mb or more RAM, 32 Mb 3D graphics card.
Platform: FS2000 FS2002 *FS-ACOF
Type: Download
Web Site: http://www.fsflighttracker.com

Title: **FS Flight Tracker**
TecPilot Rating: ✦✦✦✦✦
Description: Tired of making manual log entries of times, fuel and more? FS Flight Tracker can do it all automatically! It's much more than a flight log with fancy features. If you're a Virtial Airline pilot you won't want to make your next flight without it!.
Pub/Dev/Source: JordSoft/Keith Lovell
Rec Spec: Pentium III-1GHz or higher, 256 Mb or more RAM, 32 Mb 3D graphics card.
Platform: FS2000 FS2002 *FS-ACOF
Type: Download
Web Site: http://www.fsflighttracker.com

FS Add-On - Utilities   a-Z

Title: **FS Flightbag**
TecPilot Rating: ✦✦✦✦✦
Description: FS Flightbag is the first step in becoming an IFR pro. The GPS and other utilities included are easy to use and simple to understand.
Pub/Dev/Source: Abacus
Rec Spec: Pentium III-1GHz or higher, 256 Mb or more RAM, 32 Mb 3D graphics card.
Platform: FS98 FS2000
Type: Box
Web Site: http://www.abacuspub.com

Title: **FS Logbook**
TecPilot Rating: ✦✦✦✦✦
Description: A direct replacement for the very limited logbook functionality that is included in Flight Simulator. It will also show your flights on a map which is displayed "as the crow flies". The program can import existing logbooks, plans and FS Maintenance flight reports. It can export to ASCII (Comma Separated) to be used in many other products. This utility is very useful for Virtual Airlines as individual Flight Reports can also be generated/ exported.
Pub/Dev/Source: Lago
Rec Spec: Pentium III-1GHz or higher, 256 Mb or more RAM, 32 Mb 3D graphics card.
Platform: FS2002 *FS2004
Type: Download
Web Site: http://www.lagoonline.com

Title: **FS Maintenance**
TecPilot Rating: ✦✦✦✦✦
Description: FS Maintenance provides that "unknown" variable when maintaining aircraft. From simple Cessnas to 747-400s, it analyzes your flights and provides detailed reports on how much stress you put on the fleet - and how much your mistakes will cost you!.
Pub/Dev/Source: Lago
Rec Spec: Pentium III-1GHz or higher, 256 Mb or more RAM, 32 Mb 3D graphics card.
Platform: FS2002 *FS2004
Type: Download
Web Site: http://www.lagoonline.com

Title: **FS Meteo Weather Display**
TecPilot Rating: ✦✦✦✦✦
Description: Allows you to continually monitor the weather conditions from your aircraft and destination airport during a flight. The FS Meteo menu is not required to see weather conditions. The display operates in a similar fashion to the GPS display found on many of the aircraft that come with Flight Simulator. This packages includes a 777, 737 and Lear 45 weather displays. You must have FS Meteo installed to use this system.
Pub/Dev/Source: FSMeteo/Marc Philibert
Rec Spec: Pentium III-1GHz or higher, 256 Mb or more RAM, 32 Mb 3D graphics card.
Platform: FS98 FS2000 FS2002 *FS2004 CFS2
Type: Download
Web Site: http://www.fsmeteo.com

a-Z FS Add-On - Utilities

Title:	**FS Meteo**
TecPilot Rating:	✦✦✦✦✦
Description:	Detects where your aircraft is, reads the weather from the Internet and modifies the weather parameters in Flight Simulator. In less than one minute you have a real weather environment. A popular add-on, FS Meteo will set: Wind speed, direction, layers, gusts, aloft, temperature, altimeter, visibility, rain, snow and thunder.
Pub/Dev/Source:	FSMeteo/Marc Philibert
Rec Spec:	Pentium III-1GHz or higher, 256 Mb or more RAM, 32 Mb 3D graphics card.
Platform:	FS98 FS2000 FS2002 *FS2004 CFS2
Type:	Download
Web Site:	http://www.fsmeteo.com

Title:	**FS Panel Studio CD Version**
TecPilot Rating:	✦✦✦✦✦
Description:	FS Panel Studio can create panels for FS98, CFS, CFS2, FS2000 and 2002. Add, delete, move and resize gauges. Stretch gauge widths or heights independently. Add windows and set their positions to 1mm accuracy. When you display a panel in FS Panel Studio, it will appear almost exactly like it will in the simulator.
Pub/Dev/Source:	Flight 1
Rec Spec:	Pentium III-1GHz or higher, 256 Mb or more RAM, 32 Mb 3D graphics card.
Platform:	FS2002 *FS2004
Type:	Box
Web Site:	http://www.flight1.com

Title:	**FS Scenery Enhancer**
TecPilot Rating:	✦✦✦✦✦
Description:	FS Scenery Enhancer is a unique product that makes it very simple to add scenery elements (objects) to existing scenery. What makes this product special is that the whole process of scenery enhancing takes place inside Flight Simulator and not in a separate program. There is no code to write, no compiling needed.
Pub/Dev/Source:	Lago
Rec Spec:	Pentium III-1GHz or higher, 256 Mb or more RAM, 32 Mb 3D graphics card.
Platform:	FS2002 *FS2004
Type:	Download
Web Site:	http://www.lagoonline.com

Title:	**FS Scenery Manager**
TecPilot Rating:	✦✦✦✦✦
Description:	The FS Scenery Manager handles original and custom scenery files for Flight Simulator. Now there is no need to type in all the sceneries installed because it will automatically read all them into its database. An automatic system is provided to install a scenery from a source path or directly out of a ZIP-File.
Pub/Dev/Source:	Michael Garbers
Rec Spec:	Pentium III-1GHz or higher, 256 Mb or more RAM, 32 Mb 3D graphics card.
Platform:	FS2000 FS2002 *FS2004 CFS2
Type:	Download
Web Site:	http://www.scenery-manager.com

FS Add-On - Utilities   a-Z

Title: **FS Traffic 2002**
TecPilot Rating: ✦✦✦✦✦
Description: The next instalment of the highly popular FS Traffic series for FS2002. Brings realism to Flight Simulator by replacing default AI (Artificial intelligence) aircraft and flights.
Pub/Dev/Source: Just Flight/Lago
Rec Spec: Pentium III-1GHz or higher, 256 Mb or more RAM, 32 Mb 3D graphics card.
Platform: FS2002 *FS2004
Type: Box
Web Site: http://www.justflight.com

Title: **FS TrafficBoard**
TecPilot Rating: ✦✦✦✦✦
Description: 100% scheduled traffic at your local airport or in any of 1,800 airports with control tower and parking facilities. TTools, MyTraffic and PAI Users can now edit AI Traffic with ease. You will be surprised how easy it is to edit, generate and save AI Traffic with this utility. Manage any traffic.bgl file with the built-in WYSIWYG editor.
Pub/Dev/Source: FlightComp/Josef Dirnberger
Rec Spec: Pentium III-1GHz or higher, 256 Mb or more RAM, 32 Mb 3D graphics card.
Platform: FS2002 *FS2004
Type: Download
Web Site: http://www.simmarket.com

Title: **FS Traffic**
TecPilot Rating: ✦✦✦✦✦
Description: A powerful utility allowing you to generate traffic around any airport in the world! Comparing it with the standard dynamic traffic you will find out that the traffic generated by FSTraffic is much more powerful, customisable and will add a new dimension to your flights.
Pub/Dev/Source: Just Flight/Lago
Rec Spec: Pentium III-1GHz or higher, 256 Mb or more RAM, 32 Mb 3D graphics card.
Platform: FS98 FS2000
Type: Box
Web Site: http://www.justflight.com

Title: **FS 2002 ATC**
TecPilot Rating: ✦✦✦✦✦
Description: FS2002-ATC is an adventure creator which gives a pilot full ATC to-from anywhere in the world. A normal adventure will include: Checklists, Clearance Delivery, Ground, Cabin Announcements, Speed and Noise Restrictions, Holding Patterns, Enroute Navigation, Hand-offs, Airline Operations, IFR and VFR style flights, right down to full approach vectors for ILS and VOR approaches. Includes the ability to have non-precision approaches, like VOR and ADF and a departure setting that allows users to specify a departure and/or destination runway or just let the surface winds dictate the runway selection.
Pub/Dev/Source: Stephen Coleman

Rec Spec: Pentium III-1GHz or higher, 256 Mb or more RAM, 32 Mb 3D graphics card.
Platform: FS2000 FS2002 *FS2004
Type: Download
Web Site: http://www.stephencoleman.com/fs2000.html

a-Z   FS Add-On - Utilities

Title: FS FlightMax
TecPilot Rating: ✦✦✦✦✦
Description: FSFlightMax is modelled around the real FlightMax avionics system from Avidyne. It is a sophisticated instrument containing Moving Map, Weather Radar, TCAS Traffic, Terrain Information, Lightning Tracker and much more. It is menu driven using knobs and buttons on the device. At the time of writing this there were 18 different menus and over 22 different screen displays as well as most pages having their own configuration and sub-menu options. .
Pub/Dev/Source: Sim Systems /Avidyne/John Hnidec
Rec Spec: Pentium III-1GHz or higher, 256 Mb or more RAM, 32 Mb 3D graphics card.
Platform: FS2002 *FS2004
Type: Download
Web Site: http://www.fsflightmax.com

Title: FS LandClass (SDPs)
TecPilot Rating: ✦✦✦✦✦
Description: A huge set of global/regional ready-made Scenery Packs (SP - Used directly within FS) and Developer Packs (DP - Ready-made scenery/data used in conjunction with a utility called FSLandClass) for users and coders to enhance Flight Simulator's default geographic/topographic environment. An impressive array of 68 areas are available (at the time of writing this).
Pub/Dev/Source: Burkhard Renk
Rec Spec: Pentium III-1GHz or higher, 256 Mb or more RAM, 32 Mb 3D graphics card.
Platform: FS2002 *FS2004
Type: Download
Web Site: http://www.simmarket.com

Title: FS LandClass
TecPilot Rating: ✦✦✦✦✦
Description: FSLandClass is a tool to create new mesh textures with auto-generated scenery for FS2002 /(FS-ACOF). With the full version of the tool you can design mesh textures and auto-generated scenery in the region covered by the country packs. (see FS Landclass in Download Scenery section).
Pub/Dev/Source: Burkhard Renk
Rec Spec: Pentium III-1GHz or higher, 256 Mb or more RAM, 32 Mb 3D graphics card.
Platform: FS2002 *FS2004
Type: Download
Web Site: http://www.fssharecenter.com/fslandclass

Is there a place where I can see up-to-date listings of all these add-on products and search by their categories?

Yes of course: WWW.TECPILOT.COM
Home Of "The Good Flight Simmer's Guide"

FS Add-On - Utilities   a-Z

Title: **FS Mesh**
TecPilot Rating: ✦✦✦✦✦
Description: A new scenery design tool for MS Flight Simulator 2000 only. Intended to integrate the creation of big, high detail elevated mesh terrain and base scenery objects. Import mesh information from DEM30 - files, DHM - 100 files and FSRail files. You can enter additional information based on background images, scanned or digital maps as examples. Modify every input point to correct for errors or to adjust to your sceneries need.
Pub/Dev/Source: Burkhard Renk
Rec Spec: Pentium III-1GHz or higher, 256 Mb or more RAM, 32 Mb 3D graphics card.
Platform: FS2000
Type: Download
Web Site: http://www.simmarket.com

Title: **FS SoundScape**
TecPilot Rating: ✦✦✦✦✦
Description: FSSoundScape is a small module that has some similarity to FS Scenery Enhancer - but this module will let you add sounds to locations. Easy to use, just put your aircraft in the position where the sound needs to be heard, open the menu, select the sound you want to hear and tell the module under what condition it needs to be played. The next time you visit the location you will hear the sound.
Pub/Dev/Source: Lago
Rec Spec: Pentium III-1GHz or higher, 256 Mb or more RAM, 32 Mb 3D graphics card.
Platform: FS2002 *FS2004
Type: Download
Web Site: http://www.lagoonline.com

Title: **Final Approach**
TecPilot Rating: ✦✦✦✦✦
Description: Final Approach is a powerful, Windows-based approach chart atlas & designer for use with a range of simulators. Create your own approach charts. Use any of over 4,300 approach charts included. Print high-quality approach charts from your printer - from any chart. It does the lot!
Pub/Dev/Source: Just Flight
Rec Spec: Pentium III-1GHz or higher, 256 Mb or more RAM, 32 Mb 3D graphics card.
Platform: ALL
Type: Box
Web Site: http://www.justflight.com

Title: **Flight Deck Companion**
TecPilot Rating: ✦✦✦✦✦
Description: FDC combines the features of many chatter, and GPWS programs, adds some brand new concepts, into one audio enhancement program. FDC allows you to listen to various ATC chatter, as well as captain and co-pilot procedure call outs. FDC also features a black box, which allows you to review various parts of your flight. A ground proximity warning system is also integrated into the program.
Pub/Dev/Source: Dave March
Rec Spec: Pentium III-1GHz or higher, 256 Mb or more RAM, 32 Mb 3D graphics card.
Platform: FS2002
Type: Payware
Web Site: http://www.mypitch.co.uk/

a-Z FS Add-On - Utilities

Title: **Flight Director 99**
TecPilot Rating: ✦✦✦✦✦
Description: Integrates Real Weather, Flight Planning, GPS, EFIS, Weather Radar and ATC in the ultimate companion to Microsoft Flight Simulator 98.
Pub/Dev/Source: Just Flight
Rec Spec: 64 Mb ram. (Internet connection is required for downloading live weather data). A 3D video card capable of runing a 3D window within window.
Platform: FS98 FS2000
Type: Box
Web Site: Http://www.justflight.com

Title: **GauCleaner v3.4**
TecPilot Rating: ✦✦✦✦✦
Description: Anytime you install a panel, many so called .GAU files are copied into the GAUGES directory of Flight Simulator. When you delete the panel a consistent number of gauges still remain on your hard disk: GauCleaner will help you remove the unused gauges and files and thus keeping your HD optimised and free from unwanted files.
Pub/Dev/Source: Nicola Maragon
Rec Spec: Pentium III-1GHz or higher, 256 Mb or more RAM, 32 Mb 3D graphics card.
Platform: FS2002 *FS2004
Type: Download
Web Site: http://www.pentaonline.it/comitato_hq/gaucleaner3.html

Title: **Instant Airplane Maker**
TecPilot Rating: ✦✦✦✦✦
Description: Instant Plane Maker allows you to paint and customize your own FS aircraft. This powerful program is so simple anyone can do a custom paint job in just hours!
Pub/Dev/Source: Abacus
Rec Spec: Pentium III-1GHz or higher, 256 Mb or more RAM, 32 Mb 3D graphics card.
Platform: FS98 FS2000
Type: Box
Web Site: http://www.abacuspub.com

Title: **MyTraffic (Box)**
TecPilot Rating: ✦✦✦✦✦
Description: MyTraffic, a self-contained package which populates FS with traffic with authentic airlines. By associating airlines with particular regions, hubs and countries, they have made it possible to view airlines in the right regions. Containing several aircraft in hundreds of liveries, MyTraffic contains airlines from nearly every region across the globe. Textures have been created to minimize the impact on frame rates, finding a balance between realism and performance.
Pub/Dev/Source: Aerosoft / Burkhard Renk
Rec Spec: Pentium III-1GHz or higher, 256 Mb or more RAM, 32 Mb 3D graphics card.
Platform: FS2002 *FS2004
Type: Box Download
Web Site: http://www.aerosoft.com

FS Add-On - Utilities   a-Z

Title: **MyTraffic (Download)**
TecPilot Rating: ✦✦✦✦✦
Description: MyTraffic, a self-contained package which populates Flight Simulator with realistic traffic with authentic airlines. By associating airlines with particular regions, hubs and countries, they have made it possible to view correct airlines in the right regions. Containing several aircraft in hundreds of liveries.
Pub/Dev/Source: Burkhard Renk
Rec Spec: Pentium III-1GHz or higher, 256 Mb or more RAM, 32 Mb 3D graphics card.
Platform: FS2002 *FS2004
Type: Download
Web Site: http://www.simmarket.com/online/mytraffic/

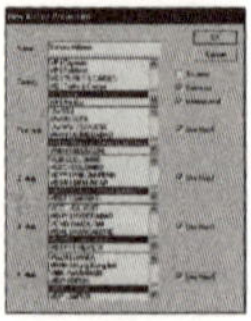

Title: **MyTraffic Editor**
TecPilot Rating: ✦✦✦✦✦
Description: MyTraffic Editor is an advanced AI Traffic configuration utility allowing the generation of thousands of flight plans at the click of a button, all based on user adjustable parameters. Capable of recognizing geographic regions and country borders, the editor makes sure that you see the right traffic in the right region. MyTraffic Editor allows you to build up your own database of planes, or to add planes that you have installed into the traffic generation.
Pub/Dev/Source: Burkhard Renk
Rec Spec: Pentium III-1GHz or higher, 256 Mb or more RAM, 32 Mb 3D graphics card.
Platform: FS2002 *FS2004
Type: Download
Web Site: http://www.simmarket.com/online/mytraffic/

Title: **NOVA - The Airport Builder**
TecPilot Rating: ✦✦✦✦✦
Description: Design 3D objects for use in Flight Simulator. Create custom hangars, control towers, buildings, bridges, terminals, gates, houses, docking systems and more. NOVA comes with an extensive library of over 250 textures, with ready made macros. This description cannot do it justice. Check out their web page for more information. We think you'll be impressed.
Pub/Dev/Source: Rafael Sanchez
Rec Spec: Pentium III-1GHz or higher, 256 Mb or more RAM, 32 Mb 3D graphics card.
Platform: FS2000 FS2002 *FS2004
Type: Download
Web Site: http://www.fsnova.com

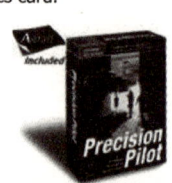

Title: **Precision Pilot**
TecPilot Rating: ✦✦✦✦✦
Description: Precision Pilot provides an interactive and graded tutorial system for pilots to learn the methodology of instrument-only navigation and associated skills. Learn what an NDB is, how to execute a holding pattern or perform a professional ILS approach.
Pub/Dev/Source: Oddsoft Ltd/Just Flight
Rec Spec: Pentium III-1GHz or higher, 256 Mb or more RAM, 32 Mb 3D graphics card.
Platform: ALL
Type: Box
Web Site: http://www.justflight.com

a-Z   FS Add-On - Utilities

Title: **Private Pilot Training**
TecPilot Rating: ✦✦✦✦✦
Description: Private Pilot Training is one of the first tools that allows you to use FS2000 as a serious flight trainer. The program gives you individualized instruction as would a real CFI. The interface for the lessons is attractive and easy to use. One program will teach you all the

Pub/Dev/Source: Abacus
Rec Spec: Pentium III-1GHz or higher, 256 Mb or more RAM, 32 Mb 3D graphics card.
Platform: FS2000
Type: Box
Web Site: http://www.abacuspub.com

Title: **Project Magenta**
TecPilot Rating: ✦✦✦✦✦
Description: A set of rich and fully functional applications, ideally run on a network, which provide glass cockpit and flight systems on dedicated PCs. Airbus, Boeing cockpit systems are available in abundance and with such accuracy! There simply isn't enough space describe all that is on offer so we suggest you visit Progect Magenta's web site for more information.

Pub/Dev/Source: Project Magenta/Enrico Schiratti and others
Rec Spec: Pentium III-1GHz or higher, 256 Mb or more RAM, 32 Mb 3D graphics card.
Platform: FS98 FS2000 FS2002 *FS2004
Type: Download
Web Site: http://www.projectmagenta.com/

Title: **Ready For Pushback**
TecPilot Rating: ✦✦✦✦✦
Description: Based on 747-200. Includes detailed fuel, hydraulic, electrical and other functional panels. You can even fuel your aircraft from the wing fuelling panel. Over 400 switches and gauges. Based on FS-2002 GMAX Models. 4 animated aircraft. Liveries include: Virgin Atlantic, Air Canada, Northwest, KLM with more available for download. Flight dynamics tested by real 747-200 Captain.

Pub/Dev/Source: Flight 1
Rec Spec: Pentium III-1GHz or higher, 256 Mb or more RAM, 32 Mb 3D graphics card.
Platform: FS2002 *FS2004
Type: Box
Web Site: http://www.flight1.com

Title: **Schiratti Control Center**
TecPilot Rating: ✦✦✦✦✦
Description: The "Swiss Army Knife" of flight-sim add-ons. This is the successor to "Schiratti Commander", with additional features and tools; a great all-rounder for hobby-pilots which is also known as "FlightZone Commander". Schiratti Control Center can graphically show you the contents of sceneries, runways, coastlines, mountains, buildings, navaids - everything. Design your own sceneries and enhance existing ones.

Pub/Dev/Source: HSP GmbH/Gerhard Kreuels
Rec Spec: Pentium III-1GHz or higher, 256 Mb or more RAM, 32 Mb 3D graphics card.
Platform: FS2000 FS2002 *FS2004
Type: Download
Web Site: http://www.hspgmbh.de/scc

FS Add-On - Utilities   a-Z

Title: **Weather Center**
TecPilot Rating: ✦✦✦✦✦
Description: A configurable, automatic weather generation application. Weather Center executes in the background while pilots fly and it constantly updates the current weather based on aircraft position relative to airports and other reporting stations in a world-wide database. The application is fully compatible with the VATSIM air traffic control network. Weather may be downloaded from any requested reporting station, edited, saved, and loaded from disk enabling offline weather simulation. .
Pub/Dev/Source: Simulation Widgets/Keith Kile
Rec Spec: Pentium III-1GHz or higher, 256 Mb or more RAM, 32 Mb 3D graphics card.
Platform: FS2000 FS2002 *FS2004
Type: Download
Web Site: http://www.simulationwidgets.com

Title: **Wetter 2002**
TecPilot Rating: ✦✦✦✦✦
Description: Creates random and real weather. Enroute weather with departure and destination airports. Weather for enroute waypoints and destination airports and start at the location of a saved flight file. World weather for pre-defined areas (Europe, Africa, America, Asia, Australia, Ocean weather - WETTER 2000 only). World weather for your own desired areas. Weather from a FS2000/2002 flight plan, following the waypoints. Changing weather, even if you stay in one area. Random rain or snow. Real weather, if you want to use METAR or SAO files.
Pub/Dev/Source: Klaus Prichatz
Rec Spec: Pentium III-1GHz or higher, 256 Mb or more RAM, 32 Mb 3D graphics card.
Platform: FS2000 FS2002 *FS2004
Type: Download
Web Site: http://home.t-online.de/home/KPrichatz/wetter.htm

SECTION thirteen

Reference

Web Sites — Reference a-Z

Air Traffic Control Developers

AITG - AI Traffic Group — http://www.aitraffic.net
Local 1 - Air Traffic Control Sim — http://www.geocities.com/hook007
Radar Contact — http://www.jdtllc.com

Aircraft Developers

Aeroplane Heaven — http://www.aeroplaneheaven.com
Aeroworks Technologies — http://www.aeroworks-technologies.com
Airliners — http://www.airliners.netfirms.com
AlliedFsGroup — http://www.alliedfsgroup.com
ARNZ - Aircraft Repaints New Zealand — http://arnz.myhost.co.nz
Boeing B 314 - The Home of The Clipper — http://b314clipper.com
Carenado — http://www.carenado.com
Flight Simulation Depot — http://www.geocities.com/fltsimdepot
FlightCraft — http://www.flightcraft.net
Freeflight Design Shop — http://www.freeflightdesign.com
Hungaryan Flight Simulator Club — http://www.fsklub.hu
International Freeware Design Group (IFDG) — http://www.4gigs.com/~ifdg
Lockheed L1011 - The Home of The TriStar — http://1011tristar.com
Mark's FSDS projects — http://www.mark-harper.co.uk/FSDS
Mirage Aircraft for Flight Simulator — http://www.mirage4fs.com
MoTIS Virtual Jet Design — http://www.motisvirtualjetdesign.com
Premier Aircraft Design — http://www.flightsimnetwork.com/premaircraft/
Project Airbus — http://www.projectairbus.com
Project XP-38N — http://www.kazoku.org/xp-38n
ProjectAI — http://www.ProjectAI.com
RealAir Simulations — http://www.realairsimulations.com
Shigeru's Aircraft Models — http://www.flightsimnetwork.com/shigeru/
Stewart-Global Aircraft — http://www.sgair.net
The Real Cockpit — http://www.therealcockpit.com
Virtual Wings — http://www.norwich.net/pstrany/virtualwings
WebinBlue — http://www.webinblue.com

Books and Magazines

Computer Pilot Magazine — http://www.computerpilot.com
FSMagazine — http://www.fsmagazine.nl
PC Pilot Magazine — http://www.pcpilot.net
TecPilot Publishing — http://www.tecpilot.com
TopSkills — http://www.topskills.com/flitsim.htm

Download Libraries

Absolute FlightSim — http://www.rickdavis.co.uk/fs/
Flightsim Zone — http://www.fsimzone.com/
FlightSims.co.uk — http://fsuk-downloads.co.uk
FS Freeware — http://www.fsfreeware.com
FS Jets — http://www.fsjets.com
FS Planet — http://www.fsplanet.com/

a-Z Reference — Web Sites

FS Videos	http://www.tecpilot.com/~la268/
FsWings	http://www.fswings.com
Sim Air 2000	http://www.simair2000.com
SimColombia	http://www.simcolombia.0catch.com
simFlight Denmark	http://www.simflight.dk
Simplanes	http://www.simplanes.com
SurClaro.com Flight Simulations	http://www.surclaro.com
The Norwegian FlightSim Hangar	http://home.online.no/~alfjoha/index.htm

Hardware Manufacturers

CH Products	http://www.chproducts.com
GoFlight Inc.	http://www.goflightinc.com
Precision Flight Controls Inc.	http://www.flypfc.com
SimKits	http://www.simkits.com
Boeing MCP747	http://www.mcp747.com
Cougar World	http://cougar.frugalsworld.com/
DuoPanel	http://home.hccnet.nl/rvdijk.mail
eDimensional	http://www.eDimensional.com
Elite Simulation Solutions	http://www.flyelite.ch
Thrustmaster	http://www.thrustmaster.com/

Maps and Charts

Desktop Wings	http://www.desktopwings.com
Final Approach Chart Atlas	http://webplaza.pt.lu/public/glorsche
FS Charts	http://www.fscharts.com
Gray Flyer	http://grayflyer.users.btopenworld.com/Indexx.html
Jeppesen	http://jeppesen.com

Miscellaneous

A320 Simulator von A.Bethke	http://a.bethke.bei.t-online.de
Airliner Business	http://www.joeyfans.com/~airbizz
AirSim	http://www.airsim.net
AOPA (UK)	http://www.aopa.co.uk
AOPA (USA)	http://www.aopa.org
Boeing 777 Simulator	http://www.boeing777.net
Capholland's Virtual Flight Services Station	http://flightsimmers.net/va/caphollands
Cockpit Company Watches	http://www.cockpitwatches.com
Flight Simulator User Group-UK	http://www.flightsimgrpuk.free-online.co.uk
FlightSim.Com	http://www.flightsim.com
FlightSims.co.uk	http://www.flightsims.co.uk
flugsimulator.de	http://www.flugsimulator.de
Flynorth	http://www.flynorth.com
FSTV	http://www.fsworld.us
Hovercontrol Helicopter Simulation	http://www.hovercontrol.com
La pagina de Jose M Gacias	http://www.telcom.es/~gacias
MSFS Gateway	http://www.msfsgateway.com
Neqinox	http://www.neqinox.com
Simcockpits	http://www.simcockpits.com
SimKits	http://www.simkits.com

The Good Flight Simmer's Guide Mk.II - 2003

Web Sites — Reference a-Z

Swedish Flight Simulator Club	http://www.swesimflight.com
The 'AirNet' Web Site	http://fly.to/AirNet/
The Boeing 737 Technical Site	http://www.b737.org.uk/
The DC-3 Hangar	http://www.douglasdc3.com
The Flight Sim Museum	http://www.migman.com
The Plane Gallery	http://www.planegallery.net/
VIDEOFLYREC	http://www.videoflyrec.com
Voo Simulado	http://pwp.netcabo.pt/ega/
World Flight Northern Ireland	http://www.wfni.co.uk
X-Plane Gateway	http://www.xplanegateway.com

Online Flying

ARTCC Argentino	http://www.artcc.com.ar
ATC-4-U 2003	http://www.geocities.com/atc4u2003
Cambridge Online Flying Club	http://www.davetidwell.com/egsc
Flight Sim Free For All Community	http://fsfreeforall.net
FSTower Live	http://www.fstower.com
Lufthans FlightSim Pilots Group	http://www.lhfspilots.com
Metro Helicopters	http://avsim.com/hangar/air/metrohelo
NVAS - InterNational Virtual Aviation Society	http://www.nvas.net
The Shetland Flyer	http://www.shetland.flyer.co.uk
VATSAM - Vatsim South America Division	http://www.vatsam.org
VATSIM Chicago ARTCC	http://www.avsim.com/hangar/satco/chicago/
International Virtual Aviation Organization	http://www.ivao.org

Online Publishing

Aero-News Network	http://www.aero-news.net
Avsim Online	http://www.avsim.com
Canadian Flightsiming	http://canada_flightsim.tripod.com/
Computer News	http://www.netspace.net.au/~johnw
Flight Simulator Nordic	http://www.fsnordic.net/
Flightservice Training Center	http://flightservice.fstower.com
FlightSim.com	http://www.flightsim.com
FlightSims.co.uk	http://www.flightsims.co.uk
flycomet	http://www28.brinkster.com/flycomet
Frugals World Of Simulations	http://www.frugalsworld.com
FSGateway	http://www.FSGateway.com
FS-Info	http://www.fs-info.nl
Jan Holmberg Flying!	http://w1.633.telia.com/~u63302866/
Mike's Flight Deck	http://www.mikesflightdeck.com
Netwings	http://www.netwings.org
Pilot's Assistant	http://www.tooby.demon.co.uk/P_Assist_Home.html
simFlight India	http://india.simflight.net
SimFlight.com	http://www.simflight.com
TecPilot Publishing	http://www.tecpilot.com
VueloArgentino On Line	http://www.vueloargentino.com.ar
Wanner Flight Simulation Site	http://www.dmarch.com/fltsim/
World of FS	http://www.worldoffs.com

a-Z Reference   Web Sites

Panel Developers

DC-3 Experience	http://www.dc3experience.com
DuoPanel	http://home.hccnet.nl/rvdijk.mail
FriendlyPanels	http://www.friendlypanels.arrakis.es
Johan's Panel Site	http://www.geocities.com/groovypanther/index.html
Ralph's Panel Shop	http://www.panelshop.com

Retail & Distribution

AvShop	http://www.avshop.com
Carprop	http://www.carprop.net
Computermate Online	http://www.compmate.com
Flight Computers Netherlands	http://www.flightcomputers.nl
Flight Sim Central	http://www.flightsimcentral.com
Luchtvaart Hobby Shop	http://www.lhshop.nl
PC Aviator (Australian Outlet)	http://www.pcaviator.com.au
PC Aviator (USA Outlet)	http://www.pcaviator.com
PC World (Simulation Page)	http://www.pcworld.co.uk/Simulation
RC Simulations	http://www.rcsimulations.com
SimMarket.com	http://www.simmarket.com

Scenery Developers

Airport for Windows Support	http://www.airportforwindows.com
Aragón en el Flight Simulator de Microsoft	http://www.sportaire.aero/simulacion/
Arnhem 2000	http://www.arnhem2000.com
Arno's FlightSim World	http://home.wanadoo.nl/arno.gerretsen
Cenareal	http://www.geocities.com/jorgepadilha/
Flightsim.Reiffer.com	http://flightsim.reiffer.com
FS Dream Factory	http://www.fsdreamfactory.com
FScene2002	http://home.quicknet.nl/qn/prive/rfaber
Getmapping.com	http://www.getmapping.com
Godzone Virtual Scenery Design	http://www.windowlight.co.nz/godzone
ISD Project	http://www.isdproject.com
SA Virtual Scenery Design	http://www.savsd.co.za
Scenery Germany	http://www.vfr-germany.de
Scenery Design Info	http://www.scenery-design.info
Scenery USA	http://www.SceneryUSA.com
SimFlyers	http://www.simflyers.net
SimWings	http://www.sim-wings.de
Steve Harding's Website	http://www.steveharding.com
SwedFlight Pro	http://www.swedflight.com
The Netherlands 2000	http://www.nl-2000.com
Visual Flight	http://www.visualflight.co.uk

The Good Flight Simmer's Guide Mk.II - 2003

Web Sites   Reference a-Z

Software Developers

AeroSim	http://www.aerosim.co.jp
FlyReal	http://www.flyreal.com
FS Flight Tracker	http://www.fsflighttracker.com
FS Nova	http://www.fsnova.com
FSMeteo Weather for FS	http://www.fsmeteo.com
Garry J. Smith	http://www.gjmith.com
geniX Software	http://genix.supereva.it
IFRPlan - Flight planning for flight simulators	http://www.jb-soft.de
MyOddWeb.com	http://www.myoddweb.com
OnCourse Software	http://www.oncourse-software.co.uk
Perfect Flight	http://www.fs2000.org
Project Magenta	http://www.projectmagenta.com
VistaMare Software	http://www.vistamaresoft.com

Software Publishing

Abacus	http://www.abacuspub.com
AeroSim	http://www.aerosim.co.jp
Aircraft Designer 2002	http://www.actpub.com
Easy Computing	http://www.easycomputing.com
ELITE Simulation Solutions	http://www.flyelite.com
Flight One Software	http://www.flight1.com
Just Flight	http://www.justflight.com
Lago	http://www.lagoonline.com

Virtual Airlines

Aegean Seagull Home	http://www.aegeanseagull.150m.com
Aero Companies 2000: Alaska Gold	http://www.aeroair.net
Aerolineas Argentinas Virtual	http://www.aerolineasvirtual.com.ar
Air Canadian	http://www.AirCanadian.com
Air England Virtual.	http://www.air-england.com
Air Jamaica Virtual	http://www.airjvirtual.com
AIR MALTA-VIRTUAL	http://amvirtual.net
Air Pacifica Airlines	http://www.airpacifica.org/
Alaska Virtual Airlines	http://www.nwva.org/nwva/
Altair Virtual Airlines	http://www.altairva.com
America West Virtual Airlines	http://www.awva.net
Atlantic Aerospace Corporation	http://www.atlanticskies.com
Big Sky Airways Inc.	http://www.bsairways.com
Bluegrass Airlines	http://www.bluegrassairlines.com
British Airways Virtual	http://www.bavirtual.co.uk
Bush Flying Unlimited	http://bfu.avsim.net
Central European VA	http://www.geocities.com/centraleuropeanva/
ConneXions Polynesia	http://www.airpacifica.org/cxp/
Continental Virtual Airlines	http://www.nwva.org/nwva/
Continental-KLM-Northwest Virtual Group	http://www.simairline.net/cnva
Czech Airlines Virtual	http://www.csav.cz
Emirates Virtual Airline	http://www.emiratesva.com/

a-Z Reference — Web Sites

EuroHarmony Virtual Airline	http://www.euroharmony.com
EuropAir	http://www.europairva.net
Evergreen International Virtual Air Cargo	http://www.eiavac.org
Fox River Helicopters & Shoreline Helicopters	http://www.bozair.bravepages.com/Foxriver/index.htm
Global Emergency Support Team	http://www.airpacifica.org/gest/
Jersey Group of Airlines	http://www.jersey-va.co.uk
JetStar International Airlines	http://www.jetstarairlines.com
KLM Royal Dutch Virtual Airlines	http://www.nwva.org/nwva/
Knight Air	http://www.knightair.org
Knight Air	http://www.knightair.org
Knight Air (PS1 Division)	http://lhr.knightair.org/ps1
Kunz Air VA	http://www.Kunzair.com
Livewire Airlines	http://www.livewireairlines.com
Mesaba Virtual Airlines	http://www.nwva.org/nwva/
Mil-Air	http://members.tripod.com/~imalm/default.htm
Northwest Virtual Airlines et al	http://www.nwva.org/nwva/
Ozark Virtual Airlines	http://www.rncspr.com/ozark
Peanuts Virtual Airline	http://www.peanutsva.com
Richmond Air	http://www.johnbmayes.com/
SimAirline.net	http://www.simairline.net
Socair Virtual Airlines	http://ww.socair.org
SUR Air	http://www.surair.net
SUR Air Virtual Airline (Buenos Aires HUB)	http://www.baires.surair.net
Swissair-Sabena Virtual	http://www.simairline.net/ssv
Tasman Pacific Virtual Airline	http://www.tasmanpacific.com
The Algeria virtual flight web site	http://www.flightalgeria.fr.st
Unity Security Force	http://www.flyunity.com/usec
Unity Virtual Aviation Community	http://www.flyunity.com
Virgin International Airways	http://www.simairline.net/via
Virtual Air 2000	http://www.geocities.com/air2kk
Virtual Armenian Airlines	http://www.geocities.com/armenianairlines
VIRTUAL Indian Airways	http://virtualairways.tripod.com
Virtual Military Airlift Command	http://vmac.us
Virtual Northwest Airlines	http://www.vnwa.com
Virtual United States Air Forces	http://www.vusafs.com
Virtual Wideroe	http://www.virtualwideroe.com
West Coast Air	http://www.flightsimnetwork.com/westcoastair/index.html

Glossary   Reference a-Z

A

A	Autotuned NAVAID
A	At or Above (constrained altitude)
AA	American
AAATS	Australian Advanced Air Traffic Services
AAC	Aeronautical Administration Communication
AAS	Advanced Automation System
AATT	Advanced Aviation Transportation Technology
ABM	Abeam
A/C	Aircraft
AC	Air Canada
ACARS	Aircraft Communication Addressing and Reporting System
ACARS	ARINC communications & Address Reporting System
ACARS	MU ACARS Management Unit
ACAS	Airborne Collision and Avoidance System
ACF	Area Control Facility
ACFS	Advanced Concepts Flight Simulator
ACK	Acknowledge
ACMS	Aircraft Condition Monitoring System
ACT	Active
ADC	Air Data Computer
ADF	Automatic Direction Finder
ADI	Attitude Director indicator
ADLP	Aircraft Data Link Processor
ADMA	Aviation Distributors and Manufacturers Association
ADS	Automatic Dependent Surveillance
AECB	Atomic Energy Control Board
AERA	Automated Enroute ATC
AFCS	Automatic Flight Control System
AFDS	Autopilot Flight Director System (also A/P F/D)
AFS	Automatic Flight System
AGATE	Advanced General Aviation Transport Experiments
AGL	Above Ground Level
AHRS	Altitude Heading Reference System
AIRS	Advanced Infrared Sounder
A/I	Anti-ice
AI	Artificial Intelligence
AL	Allegheny
ALPA	Air Line Pilots Association
ALT	Altitude
ALT	Alternate
ALTN	Alternate
ALT	HOLD Altitude Hold Mode
AM	Amplitude Modulation
AM	Aero Mexico
AMSS	Aeronautical Mobile Satellite Service
ANA	All Nippon Airways
AOA	Angle-of-Attack
AOC	Aeronautical Operation Control
AOCS	Attitude and Orbit Control System
AOM	Aircraft Operating Manual
AOPA	Aircraft Owners and Pilots Association
A/P	Autopilot
APA	Allied Pilots Association
APC	Aeronautical Passenger Communication
APMS	Automated Performance Measurement System
APPR	Approach/Approach Mode
APR	April
APRT	Airport
APU	Auxiliary Power Unit
AQP	Advanced Qualification Program
ARAC	Aviation Rulemaking Advisory Committee
ARINC	Aeronautical Radio Incorporated
ARPA	Advanced Research Projects Agency
ARR	Arrival
ARTCC	Air Route Traffic Control Center
ARTS	Automated Radar Terminal System
ASCII	US Code for Interface and Interchange
ASI	Air Speed Indicator
ASR	Airport Surveillance Radar
ASRS	Aviation Safety Reporting System
AT	At (an altitude)
A/T	Autothrottle
ATA	Air Transport Association
ATA	Actual Time of Arrival
ATC	Air Traffic Control
ATCS	Advanced Train Control Systems
ATCSCC	Air Traffic Control System Command Center
ATHR	Auto thrust System
ATIS	Automatic Terminal Information Service
ATM	Air Transportation Management
ATN	Aeronautical Telecommunications Network
ATS	Automatic Throttle System
ATSC	Air Traffic Service Communications
AUG	August
AV	Avianca
AVAIL	Available
AVHRR	Advanced Very High-Resolution Radiometer
AWACS	Airborne Warning And Control System
AWAS	Automated Weather Advisory Station
AWIPS	Advanced Weather Interactive Processing System

B

B	At or Below (constrained altitude)
BALPA	British Air Line Pilots Association
BASIS	British Airways Safety Information System
BF	MarkAir
BIT(E)	Built-In-Test (Equipment)
BRG	Bearing
BRT	Brightness

C

C	Centigrade
CAA	Civil Aviation Authority (Great Britain)
CAB	Civil Aeronautics Board
CAAC	Civil Aviation Authority of China
CAS	Calibrated (Computed) Air Speed
CASE	Computer Aided Software Engineering
CAT	Clear Air Turbulence
CAT	Category
Cat II	A Cat II approach
CBT	Computer Based Training
CDI	Course Deviation Indicator
CDU	Control display unit (Interface to the FMS)
CDTI	Cockpit Display of Traffic Information
CENA	Centred' Études de la Navigation Aérienne
CFIT	Controlled Flight Into Terrain
CG	Center of Gravity
CGS	Centimeter-gram-second
CI	Cost Index
CI	China Airlines
CIT	Compressor Inlet Temperature
CLB	Climb Detent of the Thrust Levers
CLR	Clear
CMC	Central Maintenance Computer
CNS	Communications Navigations and Surveillance
CO	Continental
COM	Cockpit Operating Manual
CON	Continuous
CO ROUTE	Company Route (also CO RTE)
COTR	Contracting Officer's Technical Representative
COTS	Commercial Off The Shelf
CP	Control Panel
CPCS	Cabin Pressure Control System
CPDLC	Controller Pilot Datalink Communications
CPU	Central Processing Unit
CRC	Cyclic Redundancy Check
CRM	Cockpit Resource Management
CRM	Crew Research Management
CRS	Course
CRT	Cathode Ray Tube
CRZ	Cruise
CSD	Constant Speed Drive
CTA	Controlled-Time of Arrival
CTA	Control Area (ICAO Term)
CTAS	Center TRACON Automation System
CTC	Centralized Train Central
CTR	Center
CTR	Civil Tilt Rotor
CTRL	Control
CVSRF	Crew-Vehicle Simulation Research Facility (NASA Ames)
CWS	Control Wheel Steering

238   The Good Flight Simmer's Guide Mk.II - 2003

a-Z Reference — Glossary

D

D	Derated
DA	Descent Advisor
DBS	Direct Broadcast Satellite
DE-TO PR	Derated Takeoff Engine Pressure Ratio
D-TO NI	Derated Takeoff Engine Fan Speed
DADC	Digital Air Data Computer
DATALINK	Digitized Information Transfer (air/ground)
DC	Direct Current Electricity
D/D	Drift Down
DEC	December
DEC	Digital Equipment Corporation
DECR	Decrement
DEL	Delete
DEP	Departure
DES	Descent
DEST	Destination
DEV	Deviation
DFDAU	Digital Flight Data Acquisition Unit
DFDR	Digital Flight Data Recorder
DFGS/C	Digital Flight Guidance System/Computer
DFW	Dallas Fort Worth International Airport
DGPS	Differential GPS
DH	Decision Height
DIR	Direct
DIR/INTC	Direct Intercept
DIS	Distance
DISCR	Discrepancy
DIST	Distance
DL	Delta
DLP	Data Link Processor
DLR	German Aerospace Research Establishment
DME	Distance Measuring Equipment
DMU	Data Management Unit
DNTKFX	DownTrack Fix
DOT	Department of Transportation
DOD	Department of Defense
DRU	Data Retrieval Unit
DSPY	Display (annunciation on CDU)
DTG	Distance-to-go

E

E	East
EADI	Electronic Attitude Director Indicator
EAS	Equivalent Airspeed
ECAM	Electronic Centralized Aircraft Monitor
ECON	Economy (minimum cost speed schedule)
ECS	Environmental Control System
E/D	End-of-Descent
EDF	Electricité de France
EEC	Electronic Engine Control
EFC	Expected Further Clearance
EFIS	Electronic Flight Instrument System
EGT	Exhaust Gas Temperature
EHSI	Electronic Horizontal Situation Indicator
EICAS	Engine Indicating Crew Alerting System
EIU	Electronic Interface Unit
ELT	Emergency Locator Transmitter
EMP	Electromagnetic Pulse
EMS	Emergency Medical Services
ENG	Engine
E/O	Engine-Out
EPR	Engine Pressure Ratio
EPROM	Erasable Programmable Read-Only Memory
EST	Estimated
ETA	EstimatedTime of Arrival
ETX	End of Transmission
EXEC	Execute

F

F	Fahrenheit
FA	Final Approach
FAA	Federal Aviation Administration
FADEC	Full Authority Digital Engine Control
FAIL FMC	Fail Failure The inability of a system
FAF	Final Approach Fix
FANS	Future Air Navigation Systems
FAR	Federal Aviation Regulations (United States)
FAR	Federal Acquisition Regulation
FAST	Final Approach Spacing Tool
FBO	Fixed Based Operator
FCC	Flight Control Computer
FCU	Flight Control Unit
F/D or(FD)	Flight Director
FDAMS	Flight Data Acquisition and Management System
FDC	Flight Data Company
FDR	Flight Data Recorder
FEATS	Future European Air Traffic System
FEB	February
FF	Fuel Flow
FGS/C	Flight Guidance System/Computer
FIR	Flight Information Region
Fix	Position in space usually on aircraft's flight plan
FL	Flight Level (FL 310 For example)
FLCH	Flight Level Change
FLIDRAS	Flight Data Replay and Analysis System
FLT	Flight
FMA	Flight Mode Annunciator
FMC	Flight Management Computer (also FMCS - FMC System)
FMGC	Flight Management Guidance Computer
FMGS	Flight Management Guidance System
FMS	Flight Management System
FO	First officer
FOQA	Flight Operations Quality Assurance
FPA	Flight Path Angle
FPA	Focal Plane Array
FPM	Feet Per Minute
FQIS	Fuel Quantity Indicating System
FR	From
FRA	Flap Retraction Altitude
FRA	Federal Railroad Administration
FREQ	Frequency
FSF	Flight Safety Foundation
FT	Feet

G

GA	Go-Around
GA	General Aviation
GAR	Go-Around
GCA	Ground-controlled Approach
GDLP	Ground Data Link Processor
GHz	Gigahertz
GMT	Greenwich MeanTime
GNSS	Global Navigation Satellite System
GPS	Global Positioning System
GPWS	Ground Proximity Warning System
GRAF	Ground Replay and Analysis Facility
GRP	Geographical Reference Points
GS	Glide Slope
GS	Ground Speed
G/S	Glideslope
GSFC	Goddard Space Flight Center
GW	Gross Weight

H

HAC	Hughes Aircraft Co
HAI	Helicopter Association International
HBARO	Barometric Altitude
HDG	Heading
HDG	SEL Heading Select
HDOT	Inertial Vertical Speed
HE	Altitude Error
HF	High Frequency
HI	High
HIRS	High-Resolution Infrared Sounder
HP	Holding Pattern
HPRES	Pressure Altitude
HSI	Horizontal Situation Indicator
HUD	Head-Up Display

The Good Flight Simmer's Guide Mk.II - 2003

Glossary — Reference a-Z

I

IA	Inspection Authorization
IAOA	Indicated Angle-of-Attack
IAS	Indicated Airspeed
ICAAS	Integrated Control in Avionics for Air Superiority
ICAO	International Civil Aviation Organization
ID	Identifier
IDENT	Identification
IEPR	Integrated Engine Pressure Ratio
IF	Intermediate Frequency
IFR	Instrument Flight Rules
IFRB	International Frequency Registration Board
IGFET	Insulated Gate Field Effect Transistor
ILS	Instrument Landing System
IMC	Instrument Meteorological Conditions
INBD	Inbound
INFO	Information
in.hg.	inches of mercury
INIT	Initialization
INR	Image Navigation and Registration
INS	Inertial Navigation System
INTC	Intercept
IPT	Integrated Product Team
IRS	Inertial Reference System
IRU	Inertial Reference Unit
ISA	International Standard Atmosphere
ISO	International Standards Organization
ITU	International Telecommunications Union

J

JAL	Japan Air Lines
JAN	January
JAR	Joint Airworthiness Regulations
JATO	Jet Assisted Takeoff
JL	Japan Air Lines
JSRA	Joint Sponsored Research Agreement
JUL	July
JUN	June

K

KG	Kilogram
kHz	kilohertz
KLM	Royal Dutch Airlines
km	Kilometer
KT	(kts) Knots
kW	Kilowatt

L

L	Left
LAT	Latitude
LAX	Identifier for Los Angeles
LCN	Local Communications Network.
LDGPS	Local DGPS
LFR	Low-frequency Radio Range
LIM	Limit
LMM	Compass locator at the middle marker
LNAV	Lateral Navigation
LO	Low
LOC	Localizer Beam
LOE	Line Oriented Evaluation
LOFT	Line Oriented Flight Training
LOM	Compass Locator at the Outer Marker
LON	Longitude
LORAN	Long Range Navigation
LOS	Line-Oriented Simulation
LRC	Long Range Cruise
LRU	Line Replaceable Unit
LVL	CHG Level Change

M

M	Mach Number
M	Manual Tuned NAVAID
MAA	Maximum Authorized IFR Altitude
MAG	Magnetic
MAINT	Maintenance
MAN	Manual
MAP	Missed Approach
MAR	March
M/ASI	Mach/Airspeed Indicator
MAX	Maximum
MAX CLB	Maximum engine thrust for two-engine climb
MAX CRZ	Maximum engine thrust for two-engine cruise
MCA	Minimum Crossing Altitude
MCDU	Multipurpose Control Display Unit
MCP	Mode Control Panel (pilots' interface to the autoflight system)
MCT	Maximum Continuous Thrust
MCW	Modulated Continuous Wave
MDA	McDonnell-Douglas Aerospace
MDA	Minimum Descent Altitude
MDL	Multipurpose Data Link
MEA	Minimum Enroute Altitude
MEL	Minimum Equipment List
MIDAS	Man-Machine Integration Design and Analysis System
MIDAS	Multi-discipline Data Analysis System
MILSPEC	Military Specifications
MIN	Minutes
MIN	Minimum
MIT	Massachusetts Institute of Technology
MLA	ManeuverLimited Altitude
MLE	Landing Gear Extended Placard Mach Number
MLS	Microwave Landing System
MMO	Mach Max Operating
MN	Magnetic North
MOA	Memorandum of Agreement
MOCA	Minimum Obstruction Clearance Altitude
MOD	Modified/Modification
Mode	Type of secondary surveillance radar (SSR) equipment.
MODIS	Moderate-resolution Imaging Spectrometer
MRA	Minimum Reception Altitude
MSG	Message
MSL	Mean Sea Level
MTBF	Mean Time Between Failures
MU	Management Unit
MWP	Meteorological Weather Processor

N

N	North
NACA	National Advisory Committee for Aeronautics
NADIN	II National Airspace Data Interchange Network II
NAS	National Airspace System
NAS	National Aircraft Standard
NASA	National Aeronautics and Space Administration
N/A	Not Applicable
NATCA	National Air Traffic Controllers Association
NAV	Navigation
NAVAID	Navigational Aid
NBAA	National Business Aircraft Association
NGATM	New Generation Air Traffic Manager
ND	Navigation Display
NDB	Nondirectional Radio Beacon
NESDIS	National Environmental Satellite
NLM	Network Loadable Module
NLR	National Research Laboratory (The Netherlands)
NM	Nautical Mile
NMC	National Meteorological Center
NOAA	National Oceanic and Atmospheric Administration
NOTAM	Notice for Airman
NOV	November
NRP	National Route Program
NTSB	National Transportation Safety Board
NW	Northwest Airlines
NWS	National Weather Service
NI	Engine Revolutions per Minute (percent)

a-Z Reference — Glossary

O

OAG	Official Airline Guide
OAT	Outside Air Temperature
OATS	Orbit and Attitude Tracking
OBTEX	Offboard Targeting Experiments
OCT	October
ODAPS	Operational OGE Data Acquisition and Patch Subsystem
OFST	Lateral Offset Active Light
OGE	Operational Ground Equipment
OIS	OGE Input Simulator
OO	SkyWest Airlines
OP	Operational
OPT	Optimum
O-QAR	Optical Quick Access Recorder
OSI	Open Sytem Interconnection
OTFP	Operational Traffic Flow Planning
OV	Overseas National Airways

P

P	Procedure-Required Tuned NAVAID
PA	Pan Am
PAR	Precision Approach Radar
PAWES	Performance Assessment and Workload Evaluation
PBD	Place Bearing/Distance (way point)
PD	Profile Descent
PDB	Performance Data Base
PDC	Pre Departure Clearance
PERF	Performance
PF	Pilot Flying
PFD	Primary Flight Display
PHARE	Program for Harmonized ATC Research in Europe
PHIBUF	Performance Buffet Limit
PHINOM	Nominal Bank Angle
PIREPS	Pilot Reports
PMS	Performance Management System
PND	Primary Navigation Display
PNF	Pilot Not Flying
POS	Position
POS	INIT Position Initialization
POS	REF Position Reference
PPI	Plan Position Indicator
PPOS	Present Position
PREV	Previous
PROC	Procedure
PROF	Profile
PROG	Progress Page on MCDU
PROV	Provisional
PS	Pacific Southwest Airways
PT	Total Pressure
PTH	Path
PVD	Plan View Display

Q

QAR	Quick Access Recorder
QNH	Quantity
QRH	The barometric pressure as reported by a particular station
QTY	Quantity
QUAD	Quadrant

R

R	Right
R	Route Tuned NAVAID
RAD	Radial
RAD	Radio
RAPS	Recovery Access Presentation System
RASCAL	Rotorcraft Air Crew Systems Concepts Airborne Laboratory
RCP	Radio Control Panel
R/C	Rate of Climb
RDP	Radar Data Processing (system)
REF	Reference
REQ	Required/Requirement
REQ	Request
RESTR	Restriction
RESYNCING	Resynchronizing
rf	radio frequency
RMPs	Radio Management Panels
RNAV	Area Navigation
RNP	Required Navigation Performance
ROUTER	ATN network layer
RTA	Required Time of Arrival
RTCA	Radio Technical Committee on Aeronautics
RTE	Route
RVR	Runway Visual Range
RW	Runway

S

S	South
SA	Situation Awareness
SAS	Scandinavian Airlines System
SAT	Static Air Temperature
SATCOM	Satellite Communications
SBIR	Small Business Innovative Research
S/C	Step Climb
SEA/TAC	Seattle/Tacoma International Airport
SEL	Selected
SEP	September
SESMA	Special Event Search and Master Analysis
SID	Standard Instrument Departure
SIGMET	Significant Meteorological Information
SITA	Société Internationale Télécommunique Aéronautique
SO	Southern Airways
SOP	Standard Operating Procedure
SOPA	Standard Operating Procedure Amplified
SP	Space
b	Speed Mode
SPS	Sensor Processing Subsystem
SQL	Structured Query Language
SRP	Selected Reference Point
SSFDR	Solid-State Flight Data Recorder
SSM	Sign Status Matrix
STAB	Stabilizer
STAR	Standard Terminal Arrival Route
STEPCLB	StepClimb
STOL	Short Takeoff and Landing
STTR	Small Business Technology Transfer Resources
SUA	Special Use Airspace
SWAP	Severe Weather Avoidance Program

T

TACAN	Tactical Air Navigation
TACH	Tachometer
TAI	Thermal Anti-Ice
TAP	Terminal Area Productivity
TAS	True Airspeed
TAT	Total AirTemperature
TATCA	Terminal Air Traffic Control Automaiton
TBD	To Be Determined
TBO	Time between Overhauls
TBS	To Be Specified
TCA	Terminal Control Area
TCAS	Traffic Alert & Collision Avoidance System
T/C or (TOC)	Top-of-Climb
T/D or (TOD)	Top-of-Descent
TDWR	Terminal Doppler Weather Radar
TEMP	Temperature
TFM	Traffic Flow Management
TGT	Target
THDG	True Heading
THR	Thrust
THR HOLD	Throttle Hold
TI	Texas International
TIAS	True Indicated Airspeed
TKE	TrackAngle Error
TMA	Traffic Management Advisor
TMC	Thrust Management Computer

The Good Flight Simmer's Guide Mk.II - 2003

Glossary   # Reference a-z

TMF	Thrust Management Function	VSCS	Voice Switching and Control System
TMU	Traffic Management Unit	VSI	Stalling Speed in a Specified Flight Configuration
TN	True North	VSO	Stalling Speed in the Landing Configuration
T/O / (TO)	Takeoff	VSTOL	Vertical or Short Takeoff and Landing
TOD	Top of Descent	VTD	Vertical Track Distance
TO EPR	Takeoff Engine Pressure Ratio	VTOL	Vertical Takeoff and Landing
TO N1	Takeoff Engine Fan Speed	V/TRK	Vertical Track
TOGA	Takeoff/Go-Around	VTR	Variable Takeoff Rating
TOT	Total	VU	Utility Speed
TRA	Thrust Reduction Altitude	VX	Speed for Best Angle of Climb
TRACON	Terminal Radar Approach Control Facility.	VY	Speed for Best Rate of Climb
TRANS	Transition	V1	Critical Engine Failure Velocity (Takeoff Speed)
TRK	Track (to a NAVAID)	V2	Takeoff Climb Velocity
TRU	True		
TSRV	Transport Systems Research Facility		
TT	Total Temperature		
TURB	Turbulence		

W

W	West
WAAS	Wide Area Augmentation System
Waypoint	Position in space usually on aircraft's flight plan
WBC	Weight and Balance Computer
WINDR	Wind Direction
WINDMG	Wind Magnitude
WPT	Way point
W/MOD	With Modification of Vertical Profile
WMSC	Weather Message Switching Center
WMSCR	Weather Message Switching Center Replacement
WO	World Airways
W/STEP	With Step Change in Altitude
WT	Weight
WX	Weather
WXR	Weather Radar

U

UA	United
UHF	Ultra-high Frequency
US	USAir
USAF	United States Air Force

X

XTK	Crosstrack (cross track error)
XY	Ryan Air

V

V	Velocity
VA	Heading to an Altitude
VA	Design Manoeuvring Speed
VAR	Variation
VAR	Volt-amps Reactive
VAR	Visual-aural Radio Range
VASI	Visual Approach Slope Indicator
VBF(LO)	Flaps up minimum buffet speed
VBFNG(HI)	High speed CAS at N g's to buffet onset
VBFNG(LO)	Low speed CAS at N g's to buffet onset
VCMAX	Active Maximum Control Speed
VCMIN	Active Minimum Control Speed
VC	Design Cruising Speed
VD	Design Diving Speed
VD	Heading to a DME distance
VF	Design Flap Speed
VFE	Flaps Extended Placard Speed
VFR	Visual Flight Rules
VFXR(R)	Flap Retraction Speed
VFXR(X)	Flap Extension Speed
VG	Ground Velocity
VGND	Ground Velocity
VH	Maximum Level-flight Speed with Continuous Power
VHF	Very-high Frequency
VHRR	Very High-Resolution Radiometer
VISSR	Visible Infrared Spin Scan Radiometer
VI	Heading to a course intercept
Vls	Lowest Selectable Airspeed
VLE	Landing Gear Extended Placard Airspeed
VLO	Maximum Landing Gear of Operating Speed
VLOF	Lift-off Speed
VM	Heading to a manual termination
VMC	Visual Meteorological Conditions
VMC	Minimum Control Speed with Critical Engine Out
VM(LO)	Minimum Maneuver Speed
VMAX	Basic Clean Aircraft Maximum CAS
VMIN	Basic Clean Aircraft Minimum CAS
VMO	Velocity Max Operating
VNAV	Vertical Navigation
VNE	Never-exceed Speed
VNO	Maximum Structural Cruising Speed
VOM	Volt-ohm-milliammeter
VOR	VHF OmniRange Navigation System
VORTAC	VHF Omni Range Radio/Tactical Air Navagation
VPATH	Vertical Path
VR	Heading to a radial
VR	Takeoff Rotation Velocity
VREF	Reference Velocity
VS	Design Speed for Maximum Gust Intensity
V/S	Vertical Speed/Vertical

Z

Z	Zulu (GMTtime)
ZFW	Zero Fuel Weight
ZNY	New York Air Route Traffic Control Center

Morse Code

A	.-	N	-.
B	-...	O	---
C	-.-.	P	.--.
D	-..	Q	--.-
E	.	R	.-.
F	..-.	S	...
G	--.	T	-
H	....	U	..-
I	..	V	...-
J	.---	W	.--
K	-.-	X	-..-
L	.-..	Y	-.--
M	--	Z	--..
0	-----	Full-stop	.-.-.-
1	.----	Comma	--..--
2	..---	Colon	---...
3	...--	Question (?)	..--..
4	....-	Apostrophe	.----.
5	.....	Hyphen	-....-
6	-....	Fraction bar	-..-.
7	--...	Brackets ()	-.--.-.
8	---..	Quotation	.-..-.
9	----.		

a-Z Reference — Index

A

AutoPilot Units *25*
Radios *25*
421C Golden Eagle *152*
737-500 for Fly! *140*
767 Pilot in Command *140*
A Bit Of Easy Theory *102*
A Century of Aviation - Lucky Lindbergh *140*
A Note From Mike Clark (Editor) *133*
A-10 Thunderbolt II *140*
A320 - Pilot In Command *141*
A320 - Professional *141*
Aarhus (EKAH) *168*
Abacus DC-3 *141*
ActiveSky wxRE *216*
Adding A Background Reference In GMAX *64*
Adding Animation To Your Models *69*
Adding The FS2002 MakeMDL Plug-In For GMAX *63*
Adding Windows, Doors And Metal Panels *72*
Advanced Hardware *11*
Advanced Simulated Radar Client *58*
Adventures *133*
Adventures Unlimited - Alitalia *134 141*
Adventures Unlimited - Azzurra Air *134 142*
Adventures Unlimited - British Airways *134 142*
Adventures Unlimited - Lufthansa *134 142*
Adventures Unlimited - Northwest *135 142*
Aerolite 103 Ultralight *143*
Air Power The Cold War *143*
Airbus Collection *143*
Airbus Professional *143*
Aircraft *139*
Aircraft Animator *216*
Aircraft Container Manager *216*
Aircraft Sound Manager *216*
Airline Flights 2000 *135*
Airline Pilot 1 *144*
Airliners Env 2002 *217*
Airport & Scenery Designer v2.1 *217*
Airport 2000 Volume 1 *144*
Airport 2000 Volume 2 *144*
Airport 2000 Volume 2 *168*
Airport 2000 Volume 3 *168*
Airport 2002 Volume 1 *144*
Airport 2002 Volume 1 *168*
Airport Scenery Basel 2002 *169*
Airport Scenery Bern 2002 *169*
Airport Scenery Lugano 2002 *169*
Airport Scenery Sion 2002 *169*
Airport Scenery Zurich 2002 *170*
Alaska Land Class *170*
Alicante 2002 *170*
Aligning Each Of The Images To Each Other *64*
Amazonian Sceneries 1 - Belem City *171*
Amsterdam Schiphol 2002 *171*
Amsterdam Schiphol Airport *171*
Angle of Attack Versus Speed *40*

Appalachians & NE 38m Terrain *171*
Applying Textures To A GMAX Model *73*
Approach to the Basic Landing *43*
Atlanta Intl (KATL) *172*
Austria Professional 2002 *172*
Austria, Swiz'd, Germany, Slovenia, Czech Rep'c *172*
Austrian Airports 1-4 *172*
A-Z - Add-On Products *133*

B

B314 Clipper *145*
Barcelona 2002 *173*
Bari Palese (LIBD) *173*
Battle of Britain Memorial Flight *145*
Beech Baron *145*
Berlin Tegel (EDDT) *173*
Big Flight Challenge ICAO Codes *131*
Birmingham (EGBB) *173*
Boeing 747 *145*
Boeing 757/767 For Fly! *146*
Boeing 777-200 Professional *146*
Boeing B314 - The Clipper (Online Version) *146*
Bornholm (EKRN) *174*
Brasilia International *174*
Bremen (EDDW) *189*
Bromma Stockholm City Airport *174*
Budapest Scenery Millennium Edition *174*
Building Aircraft & Scenery Using gMax *61*

C

Cambridge to Norwich *124*
Canada *175*
Canada LandClass *175*
Captain Speaking *135*
Captain Speaking 2002 *135*
Central & Southern England *211*
Central America *175*
Cessna 310 *146*
Changing Units In GMAX *67*
Channel Islands - Part 1 *175*
ChartsViews2 *217*
Classic Airliners 2000 *147*
Classic Eipper - Formance MX *147*
Cleaning Up *106*
Cologne Bonn (EDDK) *176*
Colouring The Fuselage *73*
Colouring The Tail Wing *72*
Combat Aces *147*
Combat Jet Trainer *147*
Combat Squadron *148*
Conclusion *28*
Concorde SST *148*
Configuring Hardware Through Menus *27*
Corporate Pilot *148*
Creating A Base Texture In Photoshop (Adobe) *71*
Creating Basic Scenery In GMAX For Flight Sim'r *76*

Index   Reference a-Z

Creating New Texture Types 74
Creating The Planes Stripe And Airline Name 71
Creating Your .mdl File In GMAX 69
Credits & Acknowledgements 6
Crosswind Landings 46
Cruise Performance 97
Custom Panel Designer 217

D

Dallas Fort Worth Intl (KDFW) 176
Dangerous Airports 176
Decent Performance 98
DeHaviland Chipmunk 148
Design Of Flaps 103
Display Boards 14
Dortmund, Munster-Osnabruck, Pader'n-Lipp't 177
Double Trouble 149
DreamFleet Cardinal 149
Dresden, Monchengladback, Lubeck, Erfurt 177
Dusseldorf (EDDL) 177

E

East and South-East England 211
East Anglia & North London - Part 4 178
Eastern Europe 178
Efficient Performance 96
Egelsbach, Bayreuth, Augsburg 178
El Hierro La Palma 2003 178
Emma Field 179
Escalante and Hurricane 179
ESSB, ESMK, ESGJ & Stockholm City 177
Eurowings Professional/Commuter Airliners 149
Eurowings Professional/Commuter Airliners 179
Executive Jets 149
Exporting The GMAX File 74
Exporting Your GMAX Model To Flight Simulator 68
EZ-Landmark 218

F

F-16 Thunderbird 150
F4U-5 Corsair 150
Fernando de Noronha 181
Final Approach 224
Finalising Textures With The Image Tool 74
Finding V1 93
Flaps Explained 101
Flight Deck Companion 224
Flight Deck II 150
Flight Director 99 225
Flight Management System (FMS) 56
Flight Plan 87
FlightDeck Companion 136
FlightSim Commander 136
Flightstar II SL/SC Series 151
Fluid Frame Rates 31

Fluidity V Frame Rates 34
Fly The Best 151
Fly The Mad Dog 136
Fly The Mad Dog 151
Fly to Hawaii: DC10 151
Fly to Hong Kong 182
Flying To Obtain Specs For Planning 89
Flying With Retractable Landing Gear 113
Frame Rates? 33
France Land Class 182
France, Andorra, Channel Islands, Benelux 182
Frankfurt 183
Frankfurt Intl (EDDF) 182
Friederichshafen, Innsbruck Altenreign 183
FS 2002 ATC 222
FS Action Scenery 218
FS Assist 2002 218
FS Clouds 2000 218
FS Design Studio v2 219
FS Flight Keeper 219
FS Flight Log 219
FS Flight Tracker 219
FS Flightbag 220
FS FlightMax 223
FS LandClass 223
FS LandClass (SDPs) 223
FS Logbook 220
FS Maintenance 220
FS Mesh 224
FS Meteo 221
FS Meteo Weather Display 220
FS Panel Studio CD Version 221
FS Scenery Enhancer 221
FS Scenery Manager 221
FS SoundScape 224
FS Traffic 222
FS Traffic 2002 222
FS TrafficBoard 222
FS2000 ATC 136
FS2002 Italy Landclass 180
FSLandClass (Scenery and Developer Packs) 180
Fuerteventura Lanzarote 2003 183
Furniture 19

G

GauCleaner v3.4 225
GB Airports 184
GB Airports 2001 183
Gear Explained 109
GeeBee Racer 152
General Aviation 137
GeoRender 1 - Flying M Ranch & Ranger Creek 184
GeoRender 2 - Bryce Canyon Airport 184
GeoRender 3 - Diamond Point 184
German Airports 1 185
German Airports 2 185
German Airports 3 185

a-Z Reference Index

German Airports 4 *186*
German Airports Selection 2002 *186*
Germany Land Class *186*
Getting Connected *55*
Glassair III *152*
Global Topographic Enhancement *187*
Glossary *236*
Goiania and Anapolis (Brazil Singles) *187*
Gothenborg (ESGG) *196*
Gothenburg Säve 2002 *187*
Gran Canaria 2003 *187*
Grand Canyon 10m Terrain *188*
Grand Canyon, Hawaii, Yellowstone, Guam *188*
Great Britain & Ireland *188*
Greatest Airliners 737-400 *152*
Greatest Airplanes: Archer! *153*
Greece & Balcans Land Class *189*
Greece, Turkey and Middle-East *189*
Guardians of the Sky *153*
Guide to Flying Online *49*

H

Hamburg (EDDH) *189*
Handling Characteristics *91*
Hannover (EDDV) *194*
Harrier Jets *153*
Harrier Jump Jet 2002 *153*
Hawaiian Islands 9.6m Terrain & Landclass *189*
Himalaja *190*
Hitting The Silicon *21*
Hong Kong *190*
Honolulu *190*
How And What To Test *90*
How Does It Work *62*
How Does Trimming Work *119*
How To Use Flaps *105*

I

Ibiza 2001 v2 *190*
Index *242*
Installing GMAX *63*
Installing The MakeMDL Plug-In *63*
Instant Airplane Maker *225*
Introduction *5*
Iron Knuckles DC-9-30 *154*
Is It Hard To Make A Model For Flight Simulator *62*
Islands of the West Indies *191*
Italy 2000 *191*
Italy, Tunisia, Malta & Corse *192*

J

Japan Land Class *192*
Jewels of the Caribbean v1 *191*
Jewels of the Caribbean v2 *191*

Jonkoping Airport *192*
Joysticks *19*
Jumbo 2003 *137*

K

KC135-R Stratotanker *154*
Keyboards - Cable V Free *15*
Kiruna 2002 *192*
Konotop AFB *193*
Korean Combat Pilot *154*
Kuala Lumpur Sepang Airport *193*

L

L-1011 Tristar *154*
La Touquet to Locarno *127*
Landing Performance *96*
Landvetter 2002 *193*
Le Monde de Chapsendor *193*
Learning to Land *39*
Legendary MiG-21 Special *155*
Leipzig (EDDP) *194*
London Airports *194*
London Gatwick (EGKK) *194*
London Heathrow (EGLL) *194*
Los Angeles Area 10M Mesh & Landclass *180*
Los Angeles Intl (KLAX) *195*
Luftwaffe Collection *155*
Luxembourg Airports *195*

M

Madeira 2002 *195*
Malaga 2003 *195*
Mallorca 2002 *196*
Malmoe (ESMS) *196*
Manaus - Eduardo Gomes Intl (SBEG) *196*
Manchester (EGCC) *196*
Manchester Intl (EGCC) *197*
Memory *33*
Menorca 2001/2 *197*
Messerschmitt 262 *156*
Mice - Cable V Free *15*
MIG-21UM Special *156*
Milan Airports 2002 *197*
Modelling A Simple House Structure *76*
Modelling A Simple Tower *76*
Modelling The Engines *66*
Modelling The Fuselage *65*
Modelling The Tail Wing *67*
Modelling The Wingspan *66*
Monitors *17*
Monument Valley 10m Terrain *197*
Mooney M20R Ovation *156*
More Advanced Peripheral Devices *19*
Mosquito Squadron *156*
M-Squared Breese 2 Series *155*

Index

Reference a-Z

M-Squared Sprint/Sport 1000 Series *155*
Multiple Monitor Set-Ups *18*
Munich (EDDM) *198*
Mustang Vs Fw 190 *157*
MyTraffic (Box) *225*
MyTraffic (Download) *226*
MyTraffic Editor *226*

N

Naming Your GMAX Objects *67*
National Parks *198*
Navigating In Winds *79*
Networking *28*
New Zealand *198*
Newark Intl (KEWR) *198*
Norwich to La Touquet *125*
NOVA - The Airport Builder *226*
Nurnberg, Stuttgart *199*

O

Operation Barbarossa *157*
Orlando Intl (KMCO) *199*
Oslo Gardermoen Airport *199*

P

P-51 Dakota Kid Mustang *157*
P-61 Black Widow *157*
Pacific Combat Pilot *158*
Pacific Northwest *199*
Pacific Theatre *158*
Painting The Wingspan And Engines *73*
Palma de Mallorca 'Son Sant Joan' Airport *200*
Pamplona 2001/2 *200*
Panel Type Add-ons *26*
Paraíba and Santa Catarina *200*
PBY-5 Catalina *158*
Pearl Harbour *158*
Pedals *22*
Performance *91*
Peru North & Lima 2002 *200*
Philadelphia Intl (KPHL) *201*
Phoenix 757-200 *159*
Phoenix Bonanza *159*
Pilot Training *58*
Piper Aztec *159*
Piper Comanche 250 *159*
Piper Cub *160*
Piper Meridian *150*
Piper Navajo *160*
Piper Seneca *160*
Placing New Objects In Flight Simulator *76*
Plotting Your Heading *81*
Porto Santo 2002 *201*
Precision Pilot *226*
Private Pilot *160*

Private Pilot Training *227*
Private Wings *161*
Processors - Recommended speeds & types *13*
Pro-Controller *57*
Project Magenta *227*
Project Magenta *28*
Putting It All Together *98*

R

Radar Contact Version 3 *137*
RAM (Random Access Memory) *14*
Ready For Pushback *161*
Ready For Pushback *227*
Real ATC/Three Real Flights *137*
Real Azores *201*
Reference *229*
Reggio Calabra Airport *201*
Reliable Landings *47*
Reno & Lake Tahoe 10M Mesh & Landclass *202*
Retractable Undercarriage *111*
Re-Using Old Models To Create New Ones *69*
Revolutionary EE\ BAC Lightning *161*
Rhein-Ruhr 2002 Delux *202*
Rio Grande Do Norte & Sul *202*
Roger Wilco *57*
Rome L.d.Vinci Intl (LIRF) *202*
Royal Air Force 2000 *161*
Royal Navy Aviation Collection *162*
Running Flight Simulator On Marginal Systems *36*

S

Sabre vs Mig *162*
San Francisco GMAX Works *203*
Santander 2001/2 *203*
Scandinavia - Sweden, Norway, Denmark *203*
Scandinavia Land Class *203*
Scandinavian Airports *204*
Scenery *167*
Scenery Airport Dübendorf *204*
Scenery Balearic Islands (Spain 1) *204*
Scenery Budapest 2002 *204*
Scenery Budapest Millennium Edition *205*
Scenery Canary Islands *205*
Scenery Greece *205*
Scenery Honolulu 2002 *205*
Scenery Spanish Airports (Spain 2) *206*
Schiratti Control Center *227*
SIAI-Marchetti SF.260 *162*
simTECH - Fokker DR1 *166*
Sky Ranch *206*
Sound Cards *17*
South Africa *206*
South East England - Part 3 *206*
South West England - Part 2 *207*
Spain & Portugal Land Class *180*
Spain, Portugal, Morocco, Canary Islands *207*

244

The Good Flight Simmer's Guide Mk.II - 2003

a-Z Reference

Index

Speakers *17*
Speed and Height *42*
Squawkbox *52*
St. Petersburg 2002 *207*
Stockholm Arlanda (ESSA) *207*
Stockholm Arlanda Intl *208*
Sturup 2002 *208*
Supermarine Spitfire *162*
SwedFlight Pro *208*
Swiss Airports *208*
Switzerland 3 *209*
TACS *56*

T

Tail-Draggers *110*
Taipei, Taiwan *209*
Tampa Intl Airport (KTPA) *181*
Target Frame Rates *35*
Tenerife & LA Gomera 2003 *209*
Terrain Puerto Rico - Virgin Islands *209*
Terramesh Europa 2002 *210*
Test Flying *90*
Texturing Introduction *70*
The 421C Golden Eagle *163*
The Big Flight Challenge *123*
The Concorde Experience *163*
The Dam Busters *163*
The Downside Of Flaps *105*
The Future *58*
The GMAX Interface *63*
The Hardest Flight Begins! *129*
The History of Online Flying *50*
The Landing Phase *44*
The Midlands - Part 5 *210*
The Piper PA28 Warrior *164*
The Processor *32*
The Rockies 38m Terrain *210*
The Three Point Landing *45*
The Tracker - Grumman S2-E/S2-A *164*
The Twin Otter *164*
The Zone *51*
Throttle Quadrants *22*
Time-Out For Reflection! *24*
TLK-39C Pilot Training Device *163*
Tornado *164*
Toronto Pearson Intl (CYYZ) *210*
Tricycle Undercarriage *111*
Trimming - Final Points *120*
Trimming Explained *117*
Trimming Simulators *118*
Triple Seven *138*
Tuskegee Fighters *165*
Twin Bonanza *165*

U

Ultralights *165*

United Airlines Flight Ops *138*
United Kingdom *211*
USB v Game Ports *15*
Utilities *215*

V

Valencia 2002 *212*
VATSIM & IVAO *51*
VFR Photographic Scenery *211*
VFR-Airfields Vol. 1 *212*
Video Cards *14*
Video Cards *34*
Vietnam Air War *165*
Virtual Airlines *57*

W

Wales & South-West England *212*
Warbrids Extreme *166*
Washington Dulles (KIAD) *213*
Weather Center *228*
Web Sites *230*
West Coast 38m Terrain *213*
Wetter 2002 *228*
What Are Polygons? *63*
What Are Primitives? *64*
What Can I Create For Flight Simulator *62*
What Is a Good Landing? *42*
What Is GMAX? *62*
What Kind Of Hardware Do I Need? *13*
What Kind Of Simmer Am I? *12*
What's Your Perspective? *24*
When Flaps Are Used *104*
When To Trim *120*
Why Do Pilots Have Difficulty Trimming? *120*
Why Is Trimming Necessary *119*
Wings Over China *166*
Wonderful Rio (Expansion) *213*
Wonderful Rio 2002 *213*
World Airliners 747-400 & 777-200 Pro *166*
World Airports *214*

Y

Yellowstone & Grand Tetons *214*
Yokes *20*

Z

Zurich Kloten Airport *214*

Other Books By TecPilot Publishing

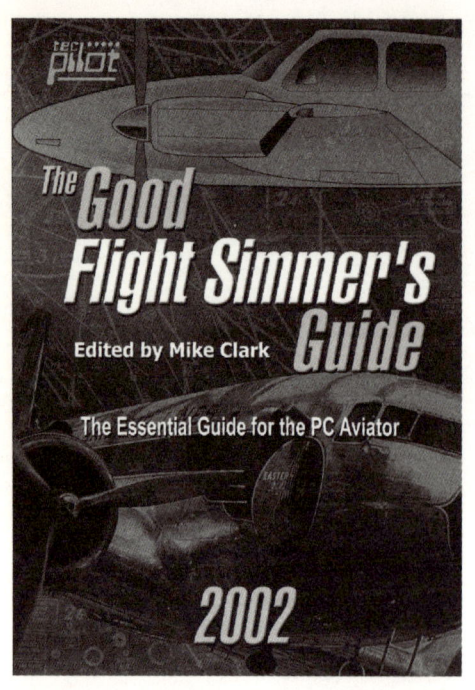

The Good Flight Simmer's Guide 2002

An introduction to the fascinating world of virtual aviation on a PC. It will inspire you to easily extend the capabilities of Flight Simulator and give you a working knowledge of how to make your flight time smooth, extremely realistic and fun. It will be a relief to many that this book is presented in a friendly and easy to follow format.

So you've flown your socks off in Flight Simulator but find that it's not as challenging or realistic as you thought it would be. You are now asking yourself if there is something you can do to update the tinny aircraft and sparsely populated scenery. You're finding that your PC simply isn't man enough for the job and renders everything like a slide show. You've heard about a world of virtual aviation where people get add-ons for FREE but you have no idea where they are or how to install them. Fear no more because this book contains ALL the answers you've been looking for - and more!

available from...

HTTP://WWW.TECPILOT.COM
or
HTTP://WWW.AMAZON.CO.UK